Under a Lucky Star

A Lifetime of Adventure

Under a Lucky Star

A Lifetime of Adventure

Roy Chapman Andrews

Foreword by Charles Gallenkamp

Afterword by Ann Bausum

Borderland Books

To Billie and George

Borderland Books Paperback Edition 2013

Copyright ©1943 Roy Chapman Andrews

Foreword and Afterword copyright ©2009 Borderland Books

Published by Borderland Books, Madison, WI
www.borderlandbooks.net

Published by arrangement with the Estate of Roy Chapman Andrews.

Printed in the United States of America

Publisher's Cataloging-in-Publication Data

Andrews, Roy Chapman, 1884–1960.

Under a lucky star: a lifetime of adventure / Roy Chapman Andrews; foreword by Charles Gallenkamp; afterword by Ann Bausum.

Originally published: New York: Viking Press, 1943.

ISBN: 978-0-9835174-3-6 (pbk)

ISBN: 978-0-9768781-8-6 (cloth)

1. Andrews, Roy Chapman, 1884–1960. 2. Naturalists—United States—Biography. 3. Explorers—United States—Biography.
4. Gobi Desert (Mongolia and China)—Discovery and exploration. 5. Central Asiatic Expeditions (1921–1930) I. Gallenkamp, Charles. II. Bausum, Ann. III. Title.

QH31.A55 A33 2009

508/.092 2008928321

Title page illustration: Steve Chappell

From the time that I can remember anything, I always intended to be an explorer, to work in a natural history museum, and to live out of doors. Actually, I never had any choice of a profession. I wanted to be an explorer and naturalist so passionately that anything else as a life work just never entered my mind.

– Roy Chapman Andrews

Contents

Foreword

by Charles Gallenkamp

The editors of *Discover* magazine have included *Under a Lucky Star* in its list of the twenty-five most outstanding science books ever written. The choice is hardly surprising. In lively and compelling prose, this autobiography of Roy Chapman Andrews traces his meteoric rise from a modest midwestern background to a wildly celebrated explorer. Much of the book focuses on Andrews's grandest venture, the legendary Central Asiatic Expeditions — a series of five daring journeys into uncharted expanses of the Gobi Desert between 1922 and 1930 that revealed a fascinating and previously unsuspected panorama of dinosaurs and extinct mammals. Indeed, many areas first explored by Andrews and his scientific team are still yielding extraordinary paleontological discoveries, and the Gobi has emerged as arguably the most prolific region on earth for Cretaceous, or late age, dinosaurs.

Organized under the auspices of the American Museum of Natural History and heavily financed by Wall Street titans and public donations, the Central Asiatic Expeditions were immense in scope and logistical complexity. Operating from a lavish headquarters — once a sprawling Manchu palace in Beijing — Andrews and his companions challenged the Gobi using automobiles supported by camel caravans, an audacious concept denounced by many skeptics as foolhardy if not impossible. In the course of their travels the explorers confronted raging sandstorms, extremes of heat and cold, political unrest, and marauding bandits, among other hazards. In retrospect, the Central Asiatic Expeditions constituted one of the truly innovative episodes in the annals of scientific discovery—an enterprise that forever changed the nature of exploration.

In an age that still idolized daredevils, soldiers of fortune, and adventurers of every kind, Andrews's role as organizer and leader of the expeditions—together with earlier escapades in Japan, Korea, and China—brought him enormous fame, making him an icon for millions of admirers and inseparably linking his name with the Gobi Desert's compelling mystique. By nature Andrews was optimistic and gregarious. His sense of humor was unfailing, his skill as a raconteur unmatched, and his enthusiasm irrepressible, though underneath these traits lurked a steely resolve that contributed to his swashbuckling, larger-than-life image. A master of diplomacy and subtle persuasion in carrying out his explorations, he was just as adept, if necessary, at forceful coercion, subterfuge, or the use of firepower. In addition, Andrews's world was a kaleidoscope of glaring contrasts: he was equally at home in the drawing rooms of New York's financiers and salons of socialites, on polo fields, at sea among hard-bitten sailors, and in camps of Mongol nomads. He was the quintessential "man of action" who devoted little time to pondering philosophical or intellectual questions. An early biographer, D. R. Barton, wrote that Andrews "possessed an insatiable appetite for kinesthetic experience. He was never merely content to see the world. He wanted to feel it—all of it—and whenever possible to wrench out whole chunks as scientifically priceless souvenirs."

Even a cursory glimpse into Andrews's life reveals an unlikely mixture of entrepreneur, scientist, adventurer, and socialite. Adored by the public and ceaselessly pursued by the press, he came as close to superstar status as any explorer of the twentieth century. Always the consummate "field man" as opposed to a desk-bound museum administrator, he was intrepid, resourceful, and courageous, as well as brilliant as an organizer and skilled as a leader. His ability as a fundraiser was extraordinary, and his genius for selecting the scientific staff—paleontologists, geologists, cartographers, archaeologists, zoologists, and support technicians—who accompanied him into Mongolia was astonishing.

When Andrews began his explorations in 1922, it was widely believed that the Gobi contained little of scientific interest. The region was generally viewed as a barren wasteland of gravel, sand, and mountains, hostile to all forms of ancient life. Andrews adamantly disagreed, saying that nobody had ever systematically searched the area. So he gambled mightily and launched his massive effort to probe the desert's secrets. The results were extraordinary. Enormous stretches along the explorers' routes were accurately mapped. Extensive studies of the Gobi's geology and ancient climate were carried out. A large collection of mammals, birds, and plants was obtained, along with a detailed photographic record of the expeditions' work.

Above all, however, it was the astounding paleontological discoveries for which the Central Asiatic Expeditions were most famous. Huge numbers of Cretaceous and Cenozoic animals were excavated, including such legendary dinosaurs as *Protoceratops* and the vicious *Velociraptor.* One site, the so-called Flaming Cliffs, was littered with dinosaur eggs, a discovery that startled and excited the entire world. Elsewhere, the explorers uncovered the gigantic bones of *Baluchitherium,* believed to have been the largest of all known land mammals. Many of the Gobi's extinct creatures were new to science, and some have revolutionized previous concepts of evolution. And as subsequent research has shown, Andrews and his companions only scratched the surface of the Gobi's treasure-trove of fossils.

Andrews considered the Central Asiatic Expeditions his crowning achievement. He hoped it would be the yardstick by which posterity would judge his success or failure in life. History has justifiably granted that wish. Long before his death, Andrews's name had become inseparably linked with the Gobi Desert, its scientific wonders, and the epic adventures of modern times. Along with such men as Peary, Scott, Amundsen, Hedin, MacMillan, and Byrd, he was one of those select few who conquered the earth's last unknown realms.

Charles Gallenkamp, writer and archaeologist, is the author of Dragon Hunter: Roy Chapman Andrews and the Central Asiatic Expeditions *and* Maya: The Riddle and Rediscovery of a Lost Civilization. *He lives in Santa Fe, New Mexico.*

Preface

Writing this book has been the means of reliving a wonderfully happy life. When I began to think it over in orderly sequence, I was amazed at the vividness of certain mental pictures. These have been the basis for the content of this book. Of course, I kept journals of my expeditions from the first in 1907 to the last in 1930. There the record is meticulously correct. But the dates of some of my personal experiences may be inaccurate. I never could remember figures and it doesn't seem important in such an informal story since I'm sure the facts are substantially correct.

Early last December, I broke my leg. It didn't seem so good at the time, but it turned out to be a lucky break. It has been a terrible winter, the worst Connecticut has known in the memory of its oldest inhabitants. An ice storm, then another ice storm and another, and blizzards! I couldn't have done much outside, anyway. In due time my wife made me comfortable on a couch in the living room before a blazing fire of fragrant hemlock logs, put a pad and pencil in my hand, and said, "Now write!" She typed the sheets as they came from my pencil. Sometimes we were completely marooned at Pondwood Farm for days, with snow piling up against the windows and the wind howling. But always there was the book. We worked all day and far into the night. I wrote "The End," on February 25, 1943. Now both of us feel lost. "What'll we do now?" asked Billie sadly. "I don't want it to end. It has been such fun."

"Don't you care," said I, "part of the story has ended but nothing else has." So if no one reads our book, it was worth doing for ourselves alone and gave us a happy winter in spite of ice and snow.

Pondwood Farm, Colebrook, Connecticut
February 25, 1943

Chapter 1

Adventure Comes Early

Often I have had to sit on a lecture platform when I was going to speak, and listen to a long introduction. It bored me stiff and likewise the audience. I wanted to get at the business in hand and the job I came for. That's the way I feel about this book. It is the story of a life in which I have had a lot of fun and excitement. So I'm going to begin the tale as quickly as possible without going back into an account of my ancestry and family. Apropos of which George Putnam quotes the remark of a visiting author to his uncle: "Major Putnam, the matter of ancestry is all very interesting. Only, the present is so much more important than the past. I always think people who are too much concerned with the pedigree of their forefathers are apt to be like potatoes—the best part of them is underground."

Of my early days there will be just enough to give a background for what follows. I don't think anyone except myself would give a tinker's damn about those boyhood years. As a rule, nothing of much interest happens to a young man until he is out of college. One chapter ought to suffice for the essential preliminaries.

I was born in Beloit, Wisconsin, at approximately two o'clock in the morning of January 26, 1884, when the temperature was thirty degrees below zero. I am told that my eyes were so slanted that when Father first looked at me he said, "Why, I've begot a Chinaman!" That remark was current in our family long before, in later life, I went to China to live for eighteen years.

Beloit lay on the banks of the Rock River in a part of southern Wisconsin that was all fields and woods and rushing streams. I was like a rabbit, happy only when I could run out of doors. To stay in the house was torture to me then, and it has been ever since. Whatever the weather, in sun or rain, calm or storm, day or night, I was outside, unless my parents almost literally locked me in.

The greatest event of my early life was when, on my ninth birthday, Father gave me a little single-barrel shotgun. Previous to this I had been allowed to shoot Grandfather's muzzle loader once or twice, but it was too much for me to negotiate with its forty-inch-long barrel.

It was with that little gun that I literally blew up my first wild goose. I was hunting just at dusk on the edge of a marsh north of Beloit. Six magnificent wild geese floating on a tiny patch of water showed up against the sunset sky. To stalk them, I had to crawl for nearly half a mile in mud and water, mostly on my stomach. Finally, I was near enough to shoot. At the roar of my gun, the three geese slowly collapsed with a gentle hissing sound, and out of a clump of bushes jumped Fred Fenton, a local sportsman. The sounds Fred made were far from gentle hissings. I'd shot his pneumatic decoys full of holes! Frightened half to death, I ran out of the marsh as fast as I could go and got home, shaking all over and sobbing hysterically. Even through my tears I could see that Father was roaring with laughter.

He kept saying over and over, "That's worth a hundred dollars to me. What I won't do to Fred Fenton! Just wait till I tell that downtown tomorrow. I'll buy you a double-barrel gun so you can get 'em all next time."

I didn't know until later that Fred Fenton was Dad's pet hate. Father was as good as his word, for the story lived in Beloit for years and I got a new double-barrel hammer gun within a week.

I sold the single gun for three dollars to a boy whose father kept a bicycle shop. That was forty-nine years ago. Only four years ago a man from Arizona wrote a letter to the "Head of the American Museum of Natural History," saying he had in his possession the first shotgun Roy Chapman Andrews ever used. It was a single barrel, still in fine condition, and he would sell it but expected a good price. What was he offered? He eventually got a reply from that same Roy Chapman Andrews, then director of the American Museum. I told him that the Museum wasn't interested, but I have always regretted that I did not buy it for myself. It was acquired by the Wisconsin Historical Society, I believe.

Even as a little boy I developed a great independence in regard to money matters. I think my allowance was ten cents a week but that didn't nearly fill my requirements for powder and shot and other things connected with shooting and fishing. So I earned my own spending money by doing odd work for the neighbors—mowing lawns, hoeing gardens, distributing circulars, raking leaves, and taking care of horses. One of the most enjoyable jobs I ever had as a boy was driving a bakery wagon when I was seventeen years old. I loved horses and my team was superb. There was a regular route which I followed, watching for the Vale Bakery signs which my customers put in their windows when they wanted me to stop.

My favorite book was *Robinson Crusoe*. My mother must have read it to me a dozen times, and I can still remember every part of it. I used to think that nothing could be more wonderful than to live on a faraway lonely island, shifting for myself.

From the time that I can remember anything, I always intended to be an explorer, to work in a natural history museum, and to live out of doors. Actually, I never had any choice of a profession. I wanted to be an explorer and naturalist so passionately that anything else as a life work just never entered my mind. Of course, I didn't know *how* I was going to do it, but I never let ways and means clutter my youthful dreams. I have often said that if I had inherited ten million dollars at birth I should have lived exactly the kind of life I have lived with no inheritance at all. A lot of money probably

wouldn't have been good for me but I believe that even independent
wealth couldn't have switched me off from exploration. The only
difference would have been that I would have financed my own
expeditions instead of getting other people to pay the bills.

My love for shooting led naturally to taxidermy. My collection
of bird skins is today in the Department of Ornithology of the
American Museum of Natural History. I taught myself how to
mount animals and deer heads by means of William T. Hornaday's
book, *Taxidermy and Home Decoration.* Learning how well I
could mount birds and animals for myself, I began to mount them
for others, and, since I was the only taxidermist in our neck o' the
woods, I had as much as I could do during the fall shooting season.
I used to have a sizable amount of money by Christmas, for every
bird and deer head shot within a radius of fifty miles came to me
if a sportsman wanted it mounted.

Beloit College was then, and is now, a fine institution. It is called
the Yale of the West, and has much the same spirit as its eastern
prototype. A boy could get a fine education if he wanted it, but,
in any event, was required to do a decent amount of work in
order to continue as a Beloit student. It was natural that I should
go there since I could live at home, and I had made up my mind
that I would finance myself through college. By summer work and
taxidermy, it would be easy, and if I earned my own money I could
spend it as I pleased.

There was no difficulty in entering Beloit College because I
came in from the "Academy" or preparatory school connected
with the institution. But once there, the question was how could I
pass freshman mathematics? All through my life, mathematics has
been my *bête noire.* Even today I add correctly only as high as I
can count on my fingers, and subtract and divide very uncertainly.
Mathematics was a required study but it was only a fog to me.
As an English student, however, I was pretty good. My English
teacher happened to be a most attractive young woman who was
in love with the man who taught me freshman mathematics. I
turned everything I had on her and on English and in the first term

was awarded frequent invitations to her house for tea and the highest marks ever given in the English department. But I flunked mathematics miserably. I indicated subtly to her that I knew I never could pass freshman mathematics and would be out of college at the end of the semester and she would lose her good student. If she could persuade her suitor to give me a passing mark in mathematics the situation would be saved. She did and it was.

I didn't do a very good job in college, so far as studies were concerned, except in literature and science. Those things came easy and I loved them. But most of the other subjects bored me exceedingly and I worked just hard enough to get medium grades. But I did a lot of other things, particularly in athletics and social activities. It was unfortunate for me that Beloit was a coeducational institution. There were too many pretty girls and I never could stay away from them. As a result I wasted a lot of time.

During my freshman year I was elected to the Alpha Zeta chapter of the Sigma Chi fraternity. Certainly nothing gave me greater happiness or greater profit than my association with this fine group of boys. It taught me to live with my fellows without selfishness or jealousy, to work for a common ideal, to have as a goal a way of life which makes for good world citizenship.

During my junior year in college came the first great tragedy of my life. A young graduate of Hamilton College, New York, had come to Beloit as an instructor in English. His name was Montague White. Monty became one of my best friends. He was handsome, a splendid athlete, and loved the outdoors as I did. Early in the spring, when ice was still in the river, Monty and I went duck shooting about seven miles north of Beloit. Our camp was on a promontory where normally a tiny creek emptied into the river, which had overflowed its banks into the low fields on the east side. On a sunny day, we paddled in our canoe to the other shore and were talking about what we would have for luncheon. I remember the decision was for corn bread and broiled wild duck.

"Well," said Monty, "we'll never get it by sitting here. Let's go."

Those were the last words he ever spoke. As the canoe shot across the river, he dropped his paddle, made a lunge to recover it, and the next second we were in the water. Monty had been in the bow. He was only ten feet from shore, where the water was not more than six feet deep, and he was an excellent swimmer, but he went down like a stone. Just the top of his head once came to the surface. The doctors said it was a stomach cramp which doubled him up like a jackknife.

I was too bewildered to realize that he was drowned for I was fighting for my own life. The racing flood whirled me along like a chip and I was fast becoming numb with cold. The current set in toward a half submerged line of willows which normally stood on the east bank. If, I thought, I can only last until I reach the trees, I've got a chance. I just made it. The rushing water forced my body against a branch and my chin hooked over another a few inches above the surface. There I hung, beating my hands until there was enough feeling so that I could draw myself half out of the water. For the next hour, climbing and swimming, I worked down the line of willows opposite the narrowest part of the overflowed fields. I was too weak to stand but the water was only a foot deep over the meadows, and on my hands and knees I crawled more than half a mile.

It was nearly wasted effort, however, for suddenly I plunged into a submerged ravine, over my head in water again. I seemed to be lying on a deliciously soft feather bed and thought how wonderful it would be just to sleep. Then a sharp pain brought me back to semi-consciousness and my head came above the surface. The barbs on a wire fence across the bottom of the ravine had ripped a great gash in my leg. That saved my life, for it stirred me into action.

"After all this I'm not going to quit now," I thought, and floundered the few yards to the other side.

A farmer and his wife saw me crawling toward their house, carried me inside, wrapped me in hot blankets, and telephoned my parents. I had been in the ice water more than three hours and was as nearly dead from cold and exhaustion as a boy could be and still live.

The police and a group of students found poor Monty's twisted body in the shallow water just a few yards from where he sank. It seemed so strange that he, the good swimmer, was the one to die almost within reach of safety while I survived against well nigh impossible odds. Still, it happened that way several times later in my life.

Although I was not physically sick after the experience, the shock and my sorrow over Monty's death played havoc with my nerves. During the next few weeks I lost forty pounds in weight and the slightest excitement would set my body to shaking and twitching painfully. All that spring and summer I spent most of the daylight hours wandering alone in the woods with field glasses and notebook studying birds and sleeping in the sun. After a year I had pretty well regained my weight and strength, but the tragedy left me with a nervous affliction from which I never recovered. Also my hair began to come out slowly and it did not stop until it was nearly gone.

A friend of mine, Hal Burr, had an amazing experience at just about that time which left him in a similar nervous condition, and we spent much time together. He was fishing on Lake Superior with his brother, Arthur, during a thunderstorm. A bolt of lightning struck Hal in the right breast and went straight through his body. The entire length of the wound was cauterized and there was no bleeding. He was helpless but conscious, while Arthur was knocked out completely although unhurt. Hal became a celebrated case in the Middle West and was exhibited at medical meetings all over the country. He recovered from all but the nervous shock and had to spend much time quietly out of doors just as I did. Neither of us could endure excitement or crowds of people. Whenever there was a baseball game or any celebration in town we both made tracks for the woods. The sound of a band or cheering would give us the jitters.

While I was in college, William Jennings Bryan spoke at Beloit, and I learned one of my first lessons in the technique of successful public speaking. All southern Wisconsin, and Beloit College particularly, was solidly Republican and so not over-friendly to so voluble a Democrat as Bryan. His speech, we had been told, was to be political and there had been much criticism of the committee for having asked him to appear. As he sat on the platform he knew, in the subtle way every lecturer does, that he faced a hostile audience. They intended to listen politely but were equally prepared to disagree with everything he said.

The president of the college introduced him. Bryan rose and stood looking out over the audience, saying nothing. He turned to the left with a broad smile and then to the right, chuckling to himself; finally he began to laugh outright but still said not a word. The audience, too, began to smile, for his mirth was infectious. Then, just at the right moment, he told a very funny story. Everybody roared and the hostility ended. His audience listened to him with sympathy and friendliness, even if without agreement. I never forgot that platform trick of Bryan's and used it to my own advantage in later years.

One of the most important events of my senior year was a visit to the college by Dr. Edmund Otis Hovey, curator of geology at the American Museum of Natural History of New York. Dr. Hovey came to lecture on the eruption of Mt. Pelée. Since the one desire of my life was to work in the American Museum, I hounded the hotel until I could see him. I spoke of my taxidermy work and very diffidently suggested that he go with me around the corner to Moran's saloon where some of my deer heads were hanging. He was exceedingly pleasant and said that he would speak to Dr. Bumpus, director of the Museum, upon his return to New York. He did so, and I wrote the director myself. A note from Dr. Bumpus replied that there were no positions open, yet if I were in New York at any time he would be glad to see me—but, of course, not to come unless I had other business in the city.

It was just the sort of polite note I, myself, as director, wrote hundreds of times in later years. But my mother and I were greatly excited at the letter. Father, more of a realist than either of us, made some uncalled for remark about counting unhatched chickens!

Graduation from college was a sad time for me. Suddenly, I realized that I had wasted time and opportunities; that although I was receiving a diploma, it really had not been earned. The graduation exercises were in the morning and that afternoon I went into the woods alone and stayed for hours, mostly sitting on the river bank. Mentally I took myself apart and examined the pieces. I didn't like what I saw. On that June afternoon I changed from an irresponsible boy to a man just as though one suit of clothes had been taken off and another put on. Then I went home, at peace with myself but consumed with a desire to get to work.

That night I told Mother of what I had been thinking. She sat there saying nothing while my words poured out in a flood. As a graduation surprise she and Father had arranged a fishing and camping trip into the great forests of northern Wisconsin. All the details were completed. She spoke about it then.

"I just don't want to go, Mother," I said. "It would be more of the same thing I've been doing. Just wasting time. I wouldn't enjoy it now. I want to go to New York and try to get into the Natural History Museum at once—next week."

She kissed me, and said, "All right, son." She was a wonderful woman, my mother.

Chapter 2

I Meet a Whale

Going to New York from Beloit for the first time was as exciting an expedition for me as any I later made into the wilds of Central Asia. Up to that time, Chicago, ninety miles away, had been the limit of my travels. The world outside our little college town literally was terra incognita and seemed a terrifyingly desirable place to a village boy. Father bought me a ticket to New York, and thirty dollars, made in the last two months by mounting birds, constituted my available capital. I spent a day in Pittsburgh and went to the Carnegie Museum, but there was no job for me there.

New York was next. I arrived at night and the magic city, as I first saw it from the deck of a Twenty-third Street ferry boat, was more beautiful than anything of which I had ever dreamed. Always I will remember how I felt as I saw it. All my fears vanished. I did not know a soul among the millions living there and yet I felt sure I should make friends and be happy; it was my city. I have never ceased to feel that way about New York.

I had been told of a small hotel on Twenty-fourth Street near Madison Avenue and went there by street car. July 5 it was, and hot as only New York can be. My room looked out on a tiny court, but even that court seemed exciting to me, with the shaded windows of other rooms seeming to beckon enticingly across it.

The next morning, Saturday, at eleven o'clock, I confronted the majestic façade of the American Museum of Natural History.

Feeling very small, I was admitted to the director's office in the east tower room, now the Osborn Research Laboratory. Years later, when I myself sat in the director's chair in that very same room, and young men and women came to see me, obviously frightened half to death, I remembered, with a tug at my heart, and tried to do for them what Dr. Bumpus did for me that day. For he couldn't have been more friendly, and my fear left me as soon as he spoke. He was tall and spare with thinning hair and a small black mustache. An air of great vitality and tremendous energy held in check dominated his personality. He seemed like a race horse at the post, ready to leap down the track at the starter's signal. He looked at me with calm appraising eyes and seemed to be making up his mind as to what sort of person I was.

We talked for some time, or rather I did, for he only sat there asking me questions. At last he said, regretfully, that there wasn't a position of any kind open in the Museum. My heart dropped into my shoes. Finally I blurted out, "I'm not asking for a position. I just want to *work* here. You have to have someone to clean the floors. Couldn't I do that?"

"But," he said, "a man with a college education doesn't want to clean floors!"

"No," I said, "not just *any* floors. But the Museum floors are different. I'll clean them and love it, if you'll let me."

His face lighted with a smile. "If that's the way you feel about it, I'll give you a chance. You'll get forty dollars a month. You can start in the department of taxidermy with James L. Clark. Now come to lunch with me. Then I'll introduce you to Clark and the others."

I even remember what we had for lunch—cold salmon and green peas at a restaurant near the Museum called the Rochelle.

There aren't many people who habitually walk faster than I do, but Dr. Bumpus was one. We raced through the Museum halls to the second floor and into the taxidermy department. In the middle of the room stood the unfinished "Carnegie Lion" given by Mr. Carnegie to the Museum; at the end was a superb polar bear and an ibex on a rock. In a corner, at a small desk, sat Jimmy Clark. He was twenty-two years old, a clean-cut, fine-looking young man. I liked him instantly. Dr. Bumpus had found him in the Gorham Silver Works in Providence, and sent him out to learn the new methods of sculpture-taxidermy from the great Carl Akeley at the Field Museum in Chicago. Jimmy had only recently returned, and the lion, polar bear, and ibex were among his first work. I was introduced, also, to Albert Butler, J. D. Figgins, and others who were making flowers and accessories. Jesse Figgins, a charming southerner, is dead after having been director of the Colorado Museum for many years. Jimmy Clark and Bert Butler are still in the American Museum in important positions.

When I left Dr. Bumpus, I called on Dr. and Mrs. W. A. Reed, formerly Beloit residents, to whom my mother had given me a letter of introduction. Fortunately, they lived near the Museum, and equally fortunately Mrs. Reed was at home. She took me to her heart at once and said she would rent me a small room on the third floor of their house for two dollars and a half a week. Thus by six o'clock Saturday night, less than twenty-four hours after coming to New York, I had a job, a room with family friends, and was ready for a new life in the place where I most wanted to be in all the world. My Lucky Star had been very active.

I had been told to report at the Museum at nine o'clock Monday morning. I reached there at eight-fifteen, but was afraid to go in that early, so I walked around the square and sat down on a granite boulder just inside the entrance to Central Park at Eighty-first Street. Looking at the impressive stone building I wondered to myself: What is it going to hold for me? What kind of friends will I make? Will I be happy there? Will I be successful? Finally, a few minutes before nine, I shut my eyes and made a little prayer, then walked to the entrance on Seventy-seventh Street and, for the first time, went through the doors of the American Museum as an employee.

A detailed chronicle of my first days in the Museum would be of interest only to myself but every moment was thrilling. My duties consisted first of mopping the floor in the morning (for Dr. Bumpus had taken my plea literally), straightening up the room, mixing clay and doing all the odd jobs of an assistant to Jimmy Clark, whom I liked better the longer I knew him.

Jimmy was possessed of great energy and ambition. He was never idle for a moment and almost lived in the Museum. Since college graduation, I had had a horror of wasting time. Thus, Jimmy's example added fuel to my inner fires. There was, moreover, no place in New York so fascinating to me as the Museum. Every Sunday, every holiday, and usually far into the night found both Jimmy and me busy on our separate interests somewhere in the building. His were mostly connected with art while mine were in study of specimens, learning photography, or reading books on natural history.

Because my existence was so completely centered in the Museum, I was able to live on my forty-dollars-a-month salary. It took pretty close figuring, however. For breakfast I had a dish of prepared cereal with milk and coffee. A bottle of milk and a box of soda crackers lasted for two luncheons. There was a little restaurant on Columbus Avenue rejoicing in the name of Savoy where meat and potatoes and coffee were served for twenty cents. This was long before the days of the automat. The Reeds often asked me to dinner and then I ate enough to last over the next day. Clothes from college were quite sufficient for more than a year.

Director Bumpus did not forget me down there in the taxidermy department. He would send for me frequently to write special labels or do some other bit of work for him. One of my first jobs was to arrange a series of caribou antlers on the walls of the stairway from the fourth to the second floor. They remain today just as I placed them thirty-six years ago. He used to inspect my floor now and then to see if the college diploma had got in the way of the mop. After eight weeks, I was given a five-dollar-a-month raise and that helped a lot.

It was about September, I think, when Jim, Bert Butler, and a young artist named Tommy Wightman and I decided to take an apartment together. It was Bert's idea and proved to be a wonderful success. The place was on One Hundred and Forty-third Street, near Lenox Avenue. Now it is in the heart of Harlem. We had a small living and dining room, two bedrooms, a kitchen, and bath. It was furnished and cost us forty-five dollars a month. Our domestic arrangements were simple. Each week one of us cooked and another did the dishes. There was an account book in which house expenditures were entered and we divided the total at the end of the month.

We had so much fun together that most evenings were spent at home. I began to have a modified social life, for the other boys knew a few girls and we went to their houses or they dined with us at the apartment. As I look back on it, I am amazed at the communal life we lived. We shared almost everything, even including our associations with girls. There was a rule that no one should bring a girl to the apartment unless it was to spend the evening with the entire group. We had a home and kept it so.

I had been in the Museum only a few months when my big chance came. In the director's office I was introduced to a fussy little gray-haired gentleman named Richardson. He was, Dr. Bumpus said, going to build a life-size model of a whale to hang in the third floor gallery-well. I was to be his assistant. I was considerably frightened but tried not to show it. What I knew about whales was less than nothing. I had never met a whale in Wisconsin's Rock River! By that time, however, I had learned to keep my mouth shut and my eyes and ears open. No one could know how ignorant I was if I didn't talk. But the job wasn't as terrifying as it sounded for we were only to enlarge a scale model which Jim Clark had made under the direction of Dr. F. A. Lucas, then director of the Brooklyn Museum.

Construction details, however, were a hidden mystery to me, for I've never had the slightest interest in mechanics. My mind doesn't run that way any more than to mathematics. But I got along all right because Richardson knew what he was about until we came to the paper covering. The framework of angle iron and bass wood strips was impressive, for the whale boasted a length of seventy-six feet. But the paper wouldn't work. It buckled and cracked and sank in between the ribs. Our whale looked awful. It seemed to be in the last stages of starvation. I used to dream about it at night, and the director was in despair.

Finally, he called Jimmy Clark and me to his office. "This whale is getting on my nerves," he said. "It is beyond all endurance. What shall we do?"

Jimmy and I knew exactly what to do for we had spent many hours discussing that emaciated whale. "Fire the paper, gentlemen," we chorused, "and let us finish it with wire netting and papier-mâché."

The director beamed. "Done. If you turn that wreck of a Cetacean into a fat, respectable whale, I'll give you both a knighthood."

Jimmy and I hopped to it with a crew of twelve men. It was amazing what a well-regulated diet of papier-mâché did for the beast. He lost that pitiful, starved, lost-on-dry-land appearance, his sides filled out and became as smooth as a rubber boot; we could almost feel him roll and blow as we built him up with our new tonic. After eight months, the job was done. During thirty-five years our whale has hung in the gallery and is still as good as new. He has been stared at by millions of eyes, and is still one of the most popular exhibits in the Museum.

Before the model was completed, my Lucky Star took charge again. A real honest-to-goodness whale was killed off the coast of Long Island at Amagansett. Jim Clark and I were sent to get the skeleton. The director's instructions, as he dashed into and out of his office, were: "Get the whole thing; photographs, measurements, baleen and skeleton—every bone." We did not learn until afterward that he never believed we could do it. He and Dr. Lucas knew a lot about beached whales and how quickly the great bones sink into the sand.

I was the most excited and the proudest boy in New York State as we journeyed toward Amagansett. Only seven months in the Museum and off on an expedition! True, it wasn't an expedition to the Arctic or the tropics, but it was an expedition nonetheless.

Once we had arrived at the village, the business of buying the whale was quickly done. The baleen, or "whale bone," was the valuable part, for at that time it was still being used for corsets and carriage whips. I believe it cost us thirty-two hundred dollars, which was only a little more than the commercial value. They threw in the skeleton, but we were obligated to get the bones ourselves.

The carcass was beached just at the edge of low tide. After the fishermen had stripped off the blubber, they went away. Jimmy and I were faced with a real problem, for the skeleton lay embedded in some fifty tons of flesh. Of course we could do nothing alone, and the fishermen were not at all eager to work even for high wages. The thermometer stood at twenty above zero and the wind was bitter.

Finally, we did persuade half a dozen men to hack away at the carcass with great knives. A horse helped to drag off huge chunks of meat by means of ropes and hooks. It was a slow business, but at last the head was separated and on the beach; also the ribs of the upper side. Then the worst happened. A storm blew up from the east, beating upon the exposed coast with hurricane force. We saw it coming and anchored our whale as best we could, working hip deep in the icy water.

For three days the shore was a smother of white surf. Anxiously we waited. Only half the skeleton was on the beach and that would be almost worthless if the remainder were lost. The fourth day was dead calm but very cold; twelve degrees above zero at noon. When we got to the beach, a smooth expanse of sand, innocent of whale, met our eyes. The bones had disappeared!

Jimmy and I were frantic, but the anchor ropes extended down into the sand where the bones had been. A little shoveling exposed the skeleton, deeply buried. It would have been difficult enough in the best circumstances to uncouple the huge vertebrae and get the ribs of the lower side, but now it was almost impossible. As soon as we dug out a shovelful of sand to get at a bone, the depression filled with water. We had to grope blindly with small knives, our arms in the freezing water up to the elbows, to disarticulate each vertebra.

Jimmy and I carried on alone for three days, warming our hands every few minutes over a driftwood fire. It seemed hopeless, and I don't mind saying that I never have suffered more in any experience of my life than I did then. But the director had told us to get every bone, and we simply couldn't give up. At last, some of the fishermen decided to help when they saw us two kids struggling hopelessly in that icy water. I believe it was more shame than the high wages which brought them to our assistance. Anyway, half a dozen men came, and we began to make real progress. At the end of a week a huge pile of bones lay well up on the beach. We checked them off one by one on a drawing from a skeleton in the Smithsonian Institution. They were all there except the pelvic rudiments and eventually these were rescued from the pot where the blubber was being tried out for oil. That meant we could return to New York with a clean bill of bones.

Those first months in the Museum were, I believe, the happiest of my life. Dr. Bumpus had given me not only his complete trust but his friendship as well. He took every opportunity to introduce me to visitors of importance and often asked me to walk with them through the Museum.

The man whom I most wanted to meet in the entire world was Frank M. Chapman. His *Handbook of Birds of Eastern North America* had been my Bible. He was on an expedition to the Canadian Rockies to collect a group of ptarmigan with the famous bird artist Louis Fuertes when I came to New York. One day, a wave of excitement ran through the building. Frank Chapman had returned. He came with Dr. Bumpus to the taxidermy department to inspect some of his newly arrived specimens. When I was introduced I could hardly speak. To him I was only an embarrassed young man; to me, he was the living embodiment of everything I wanted to be. Frank Chapman is now seventy-nine years old but he still brings with him a wave of something fresh and vital whenever he enters a room.

The president of the Museum's board of trustees, when I came there, was Morris K. Jesup, the great railroad builder. An extraordinary man he was. With virtually no formal schooling, he had amassed a large fortune and used it to give to others the education he had been denied in his youth. He was rather a terrifying figure, as I remember him, but probably my estimate wasn't a true one for all those great financiers whom I met in my youthful days seemed frightening although they were always kind to me.

Mr. Jesup used to lunch at the Museum two or three times a week. How great a place he thought it was is evidenced by the fact that he willed virtually all his fortune, about six million dollars, to the American Museum, the income to be used for research and exploration.

I met Andrew Carnegie, too, for a few moments one June day in the taxidermy department with the director. He came bouncing in to see the "Carnegie Lion" which Jimmy Clark had modeled. That was the impression he gave—bouncing. A small man with a white goatee, he seemed to radiate restless energy. He looked at the lion and said appreciative words but his eyes were roving over the entire room and, in almost bird-like movements, he flitted from one place to another, asking questions. I'll wager he knew everything that went on in that department within fifteen minutes.

Dr. Frederick A. Lucas was at that time director of the Brooklyn Museum of Arts and Sciences. In the course of building our whale model, which was done under his direction, I had seen him often. One day he said to me:

"I want someone to understudy me as director, an assistant who can take my place, eventually. I believe you would fit the job admirably. The beginning salary will be eighteen hundred dollars a year with increases as you get into the work. How would you like to come to Brooklyn?"

I didn't know what to say. The salary of one hundred and fifty dollars a month was princely for I was getting only sixty-five dollars. Yet I didn't want to leave the American Museum or go to Brooklyn. Moreover, I wasn't working for the money in my job. On the very same day a telegram came from Washington asking if I would accept a commission as a second lieutenant in the Philippine

Scouts. With a lot of other boys, I had filled out a questionnaire for the Scouts before I left college.

Having two attractive opportunities suddenly thrown into my lap put me into a complete dither. It didn't take long to dismiss the Scouts, however, for I was getting plenty of excitement right in New York. But the Brooklyn offer was a different matter. I talked it over with several men in the department. They all said I was a fool not to accept; that such an offer might never come again. True enough, but they were thinking only of the present. I knew that if I went to Brooklyn, I'd stay *inside* the institution as an administrator. What I wanted was scientific exploration in the far places of the earth, even though I loved museum work. Dr. Bumpus wished me to continue in the American Museum. He offered no increase in salary for he said he didn't want me to stay if money was the object. I would get an increase in salary as I earned it.

Looking back over thirty-six years, I believe my decision to remain in the American Museum was perhaps the wisest and most important I ever made. It was one of the major "stop lights" when one has to choose between the long straight road or an attractive bypath.

Chapter 3

Submarine Courtship

Mr. George S. Bowdoin, one of the partners of J. P. Morgan and Company and a trustee of the American Museum, was financially responsible for the whale model Jimmy Clark and I had built. In those days, and, as a matter of fact, until the Depression of 1930, being a trustee of a public institution was an expensive honor. I never shall forget a meeting of our trustees in the early part of the century at which I happened to be present. A hundred and twenty-five thousand dollar deficit in the budget for the ensuing year had been presented. Nothing was said about it until the end of a very delicious dinner with vintage champagne, when Mr. Morgan got to his feet.

"We've got to do something about this damned deficit," he said. "I'll give ten thousand. Cleve Dodge, you'll give another ten thousand. Arthur James, I'll put you down for ten thousand." And so it went around the table. Mr. Morgan didn't ask them, he told them, and there wasn't a voice raised in protest although some of the men smiled rather ruefully. In less than ten minutes the hundred and twenty-five thousand dollars was subscribed. "Now," said Mr. Morgan, "everybody's happy so let's enjoy the evening."

When I was director of the Museum from 1934 until 1941 I used to think back with longing to those happy days. The sons of the same men were our trustees but the world and their fortunes had sadly changed. Where their fathers could give ten thousand dollars with hardly a wince, a thousand was a sum not to be donated without due thought by them. Often I have presented a budget with a fifty-thousand-dollar deficit and have had the board regard me as some evil-smelling object that had suddenly been thrown into a pleasant gathering.

It was Mr. Jesup, I think, who had told Mr. Bowdoin that a collection of whales was a necessity for the American Museum and that it was up to him to do something about it. Apparently whales were about the last thing Mr. Bowdoin wanted to do anything about, but he had come across like a little man to the tune of ten thousand dollars. He was, however, frightfully embarrassed and seemed to be afraid of being laughed at by his friends who were giving Rembrandts and Van Dycks to the Metropolitan Museum of Art. I overheard him remark to the director one day:

"When Morgan comes back from Europe he'll say, 'My God, George, what in hell do you want with whales?'"

But Mr. Bowdoin's ten thousand dollars was God's gift to an enthusiastic young man even though it did embarrass the donor. It not only made possible the whale model on which I had won my spurs, so to speak, but the Amagansett whale as well. Also it launched me on a career of blubber and brine which lasted for eight years and carried me twice around the world.

I had become so keen about whales that after the Amagansett skeleton was cleaned, I begged the director to be allowed to describe it. This was my first scientific paper and when it came off the press I was as proud as though I'd given birth to quintuplets. In describing the skeleton it was necessary to use a lot of impressive technical language, and so I felt that I had become a real scientist at last. It was a pretty good paper at that, for although these particular whales had been killed for centuries, no trained scientist ever had been on the spot when a specimen was fresh. Study of the whale literature also made it clear that almost nothing was known about the habits of the beasts. Whalemen knew how to kill them and had described that well enough. But when it came to really accurate knowledge of their life histories, the books were full of fantastic yarns. There never was a more virgin field for an enthusiastic young naturalist.

Three shore whaling stations had been established on the west coast of Vancouver Island and southeastern Alaska. The great beasts were hunted from little ninety-foot steamers, killed with an explosive harpoon, towed to the factories on shore and there converted into oil and fertilizer. Knowing that some of Mr. Bowdoin's ten thousand dollars was left, I figured I might use it to advantage if the director could be made to see it my way. It didn't take much persuading when I told Dr. Bumpus I was willing to go without salary. If he would risk a possible thousand dollars of Mr. Bowdoin's money I'd guarantee to bring him some results—or else!

Thus it came about that in the spring of 1908 I set out on my first real expedition equipped with notebooks, field glasses, tapes and measuring lists, and, most of all, a Graflex camera. That instrument represented the winter's saving from my salary, which had been increased to one hundred dollars a month. This put me in the big money. Jimmy Clark was getting ready to go to Africa as a volunteer assistant to the English naturalist-photographer A. Radcliff Dugmore, and we had worked on photography all winter.

Victoria, Vancouver Island, was where I had to sell my plans and myself to the Pacific Whaling Company. Victoria slept peacefully in the sunshine while the spring days drifted into summer. About it there was a delightful air of being on the edge of a great wilderness, which, indeed, it was. At that time parts of the island's interior were almost unexplored, and one saw in the shops magnificent wapiti and deer heads, bear skins, totem poles, and baskets from the coastal Siwash Indians.

Twenty miles up the great gash which Barclay Sound cuts into the west side of Vancouver Island, the whaling factory huddled against a fir-clad mountain. I came to it on a little coastal steamer, weak from seasickness. We had battled a terrific storm all night and I had my first experience with what was to become a major problem during all my whaling days. Eventually I recovered permanently and now no storm makes the slightest difference to my semicircular canals, but with the irony of fate it came too late. In thinking back over a somewhat adventurous life, my considered opinion is that nothing I have ever done required more unadulterated guts than going out on those tossing, twisting whaling vessels with the certain knowledge that I would suffer the tortures of the damned. Often I was so weak that I lay on the deck behind the harpoon gun like a dead thing and only when one of the sailors lifted me to my feet and hooked my arm about a stay could I work the camera and take my notes.

But dogged persistence, of which, God knows, I now would be incapable, drove me out on the whaling ships to do my work. With nothing to distract my mind except a heaving stomach, I could time how long whales remained at the surface when feeding, how often they blew, how they used their flippers and flukes and a hundred other details which no whale hunter had time to note in the heat of battle.

Never will I forget the first time a whale came up beside the ship. The water was without a ripple and I saw it rising twenty feet below the surface. A huge ghostly shape driving swiftly upward until it erupted like a submarine volcano almost beneath my feet. I could look down into the cavernous nostrils as they swelled out and shot a geyser of hot stinking vapor into my face. A second later the crash of the harpoon gun sent me reeling backward. Crawling to the rail I had a momentary glimpse of a great tail rising and falling in a smashing blow that could have crushed our little boat to splinters. Then dead quiet except for the slow rattle of the rope as the lifeless body sank straight down into the blue depths. Several times it happened before I could steel my nerves to think and act. Then with my face pressed into the camera hood I waited to press the button till the phantom body showed in the mirror.

There was always the fascination of wondering what went on beneath the surface when we hunted a school of whales. Often I saw them feeding quietly and then, at some signal, heave their vast bulks into the air and disappear at the same instant. They might double and reappear behind the ship breaking water in line like a company of soldiers marching on parade. How did they communicate with each other? It is not given us to know and I doubt if it ever will be. How, too, can they descend to depths where the water pressure would crush the steel sides of a submarine like an eggshell and return to the surface a few moments later with the rush of a leaping salmon? Too rapid emergence is the terror of the deep water diver, else his body is twisted and racked by the deadly "bends." Something like eight hours for decompression is, I believe, required for a dive of three hundred feet; or, at least, it was before my adventurous friend Captain John Craig showed that helium could be mixed with oxygen and cut the decompression period amazingly.

It was positively indecent the way I pried into the private lives of whales during those days at sea. With field glasses from the masthead, I watched the love-making of a pair of humpback whales fifty feet long. An amorous bull whale may be very amusing to us but to his lady friend he is doubtless as exciting as a matinee idol is to a debutante. In this particular case the gentleman whale executed a series of acrobatic performances evidently with the object of impressing the female. He stood on his head with the tail and fifteen feet of body out of water. The great flukes were waved slowly at first; then faster until the water was pounded into spray and the terrific slaps on the surface could be heard a mile away. This performance ended, he slid up close to the female, rolling about and stroking her with his right flipper. She lay on her side apparently enjoying his caresses. Then he backed off and dived. I thought he had left her for good but she lay quietly at the surface; she knew full well that he would not desert her—yet. He was gone for, perhaps, four minutes, then with a terrific rush he burst from the water throwing his entire fifty foot body straight up into the air. It was a magnificent effort and I was proud of him. Falling back in a cloud of spray he rolled over and over up to his mate, clasping her with both flippers. Both whales lay at the surface, blowing slowly, exhausted with emotion.

I felt embarrassed to be spying on their love-making like a Peeping Tom but the captain was made of sterner stuff. The exhibition left him cold. His materialistic mind visualized the thousands of dollars their carcasses would bring in oil and fertilizer to the exclusion of all else. From the mast head I pleaded with him to "have a heart" but without avail. The ship slid closer and closer to the half slumbering lovers and a bomb-harpoon crashed into the side of the amorous bull. Half an hour later the lady, too, was killed for she refused to leave the vicinity of her dead lord.

I suppose I am the only naturalist who has ever been present at the birth of a baby whale. One day a big female finback was brought to the station, obviously in an "interesting condition." The captain told me he had killed her only a few miles from shore where she was probably seeking quiet water for the accouchement. Wire cables were made fast about her flukes and as the steam-winch drew the sixty-ton body out of the water the baby was born right before our eyes. It was twenty-two feet long and would weigh about fifteen tons; the mother measured sixty-five feet. Of course the reason why the babies are so proportionately large is because whales live in a supporting medium.

Milk was oozing from the teats of the whale and I drew off a pailful just like milking a cow. The taste was not good but it was so strongly impregnated with the gases of decomposition that not much of the original flavor was left.

Absolutely nothing was known about the breeding habits of whales so I went into the study in a big way. The uterus of an eighty foot sulphurbottom is about as large as a double bed, and hunting for an embryo is a messy job. By recording the size of each fetus and the dates, it became evident that there is no regular breeding season for whales although the "joys of spring" seem to have a slight effect upon their love making.

Some years later the British government became worried about the rate at which whales were being killed in the South Atlantic. They wanted to enact laws to protect them during the breeding season. Would I please tell them when whales bred? It seemed as simple as all that but my records of pregnant females showed that they couldn't impose blue laws or birth control on a bull whale. By the same token if they put a ban on the human hunters no good would come of it. We can never know whether or not whales have any constancy in their marital relations. But with the whole ocean to roam in, I should judge that free love was the order of their lives. Still, their great hearts, as big as an office safe, do know the tender feelings of affection—at least mother love, as I have seen many times.

How large whales nurse their young is still a mystery. The two teats on either side of the genital opening are only about two inches long. The young whale's lips are six or seven inches thick and the snout is pointed. Unless the babies nurse with the teats above the surface it seems inevitable that more sea water than milk would enter their stomachs.

These new facts of life history and many more were learned during my stay at the whaling station in Vancouver Island and Alaska. A hundred large whales of three species were measured, photographed, and described, and by the time I returned to New York more new data had been accumulated in my notebooks than had ever been gathered about any group of water mammals in the same length of time. The reason, of course, was because I happened to be the first naturalist to study whales at sea with the unequalled opportunities which shore stations presented. Moreover, at the Barclay Sound factory I had roughly cleaned and prepared for shipment the skeleton of a humpback whale, so the trip represented an actual specimen as well as factual data.

Chapter 4

Muscles and Murderers

Returning to New York in the autumn of 1908 was an event to me. Suddenly, in the Museum, I had become something more than just a young employee. The photographs of living whales created a mild sensation in the newspapers, and for the first time I saw my name displayed in the *New York Times* and other metropolitan dailies. Mine were the only pictures ever taken, up to that time, of the biggest and least known group of mammals in the world.

A representative of the monthly magazine *World's Work*, edited by Mr. Walter Hines Page, asked me to write an article on modern whaling, illustrated with my best photographs. For it they would pay two hundred and fifty dollars. That was a lot of money in those days. I did the article and sent it to Mr. Page. His son, Arthur, now vice-president of the American Telephone and Telegraph Company, came to see me and suggested that I dine with him and his father at their apartment. He didn't say anything about the article. Arthur was about my own age, and I liked him at once. The dynamics of his personality had been directed into channels different from mine but no less exciting because they concerned literature and public affairs.

His tall, gangling father seemed to envelop me in a warm smile the moment I looked into his eyes and grasped his hand. I felt that I was talking to a great man; how great a man the world learned in the war-torn years when he was Ambassador from the United

States to the Court of St. James, the most important diplomatic post in the world.

We had a delightful dinner and after it Mr. Page asked me to come into his study. My article was lying on the desk.

"This," said he, "has the basis of a fascinating story, but you have so obscured it with technical language that even I can read it only with difficulty. The general public could get nothing out of it; they wouldn't bother to read beyond the first sentence. After all, what are you writing it for? Not just for the money *The World's Work* pays you! You want it to be read. You want to share the things you did and saw with the greatest number of people. Now, if you'll just forget that you know a word of scientific terminology and write this story just as you have been telling it tonight to Arthur and me, you'll really be doing a service."

I took the article back and rewrote it as nearly as possible the way I had talked that night. Mr. Page was pleased and *The World's Work* ran it as the lead with a score of photographs. That talk, and others I had with Mr. Page, were of paramount importance to me. I realized before long that it was fun to relate my experiences but that I'd never be a great writer; that I couldn't produce literature. All I could do was to set down whatever story I had to tell as simply as possible. That is all I have tried to do ever since.

My whale photographs, made into colored lantern slides, were unusual and a few really spectacular. I was invited to give an illustrated lecture at the New York Academy of Sciences on my discoveries in the life history of whales. The half hour's speech I wrote out, learned by heart, and promptly forgot. After a few awful moments of floundering, there was nothing to do but go on with the story extemporaneously, using the lantern slides as notes. It was surprisingly easy and I resolved then that if I couldn't learn to think and talk on my feet, I'd never speak publicly at all.

Just before leaving for Vancouver and Alaska, I had been transferred to the department of mammals and birds as assistant to Dr. J. A. Allen. Frank M. Chapman was his associate curator, in charge of birds, while Dr. Allen devoted himself to mammals. Dr.

Allen, who lived to ninety-one, and worked in the Museum almost until the day of his death, was the dean of the pioneer naturalists. No sweeter character or truer scientist ever lived. The days I spent at his side were all too few, but to work with him was one of the greatest privileges of my life.

That autumn of 1908 I entered Columbia University to study for the degree of Doctor of Philosophy. During the mornings I worked at the Museum. One of my most interesting jobs was revising and identifying the collection of rabbits. What I didn't know about rabbits by the time that job had ended was nobody's business. Just by looking at a rabbit I could tell you where he came from, what his ancestry was, and even his prenatal influences. Squirrels came next and I delved into their private lives in the same uninhibited way. Then seals. White-footed mice were also on the list because Wilfred H. Osgood, of the Biological Survey in Washington, had just published a monumental work on the genus *Peromyscus*— wood mice to you!

In the afternoon, I had the real opportunity of studying under Professor Henry Fairfield Osborn when he gave his last series of lectures as Da Costa Professor of Zoology at Columbia University. The class was the "Evolution of Mammals" and his assistant, Dr. William King Gregory, put us through our paces for an hour of preliminary work at every session. The class met at two o'clock in Schermerhorn Hall at the university. After Will Gregory had our mental machinery running in high gear Professor Osborn took over for two hours. Then we had a rest, all the windows were opened to give us a spot of New York's carbon-monoxide, and then was another hour digesting what Professor Osborn had said, under the direction of Will Gregory. By that time, all of us were feeling the need of something more substantial than fossil bones, so usually we walked, or took the subway, to Heine's saloon at Eighty-first Street and Columbus Avenue. There, in the back room, at a table presided over by "George," who served us admirable dark beer and a passable dinner, we continued to argue over the subject of our lectures far into the night.

There were only eight in the class including one very attractive girl. She usually dropped out before we went to Heine's because that place wasn't considered *au fait* for women in those days. Anyway we were glad to have her leave, for our discussions as to whys and wherefores of dinosaurs and pterodactyls could be much less restrained.

Roy Waldo Miner, Curator Emeritus of Invertebrate Zoology in the Museum, was of our number. Another was a visiting scientist from Cambridge University, England, who had come over to take Professor Osborn's course. He was C. Forster Cooper, now director of the British Museum of Natural History, London. We had another most stimulating fellow-student. He was a pugilistic chap, brilliant as a meteor and always ready to take the opposite side of an argument whether or not he believed in what he said. I remember that when he took the oral examination for his degree, one of his questioners said something that set him off. He started an argument that had nothing whatever to do with the defense of his thesis. When he was called up sharply, he yelled that if getting a doctor's degree at Columbia University would make him such a stuffed shirt as a lot of those in the room he didn't want it and they could all go to hell. The examination adjourned abruptly and for a week his degree hung in the balance. Finally some of his professors who understood his volatile temperament persuaded the outraged members of the faculty that they couldn't withhold the degree from a man of such brilliance, just because he had lost his temper.

Heine's was the luncheon place of the Museum staff for several years. We had a table in the center of the room at which anything might happen. I remember one day Louis Agassiz Fuertes, the bird artist, and Ernest Thompson Seton got into an argument as to who could give the best imitation of a coyote's howl. Both lifted their voices in the prairie song which brought two policemen on the double quick. Just as the cops were entering the door, Fuertes began squawking like a macaw. George told the police it was only the zoologists from the Museum and not to mind them—it happened most any day. Better they should have a glass of beer and enjoy the show!

The afternoons, when not at the university, I worked in the graduate's dissecting room at the College of Physicians and Surgeons, studying comparative anatomy under the brilliant Dr. George S. Huntington. At that time "P. and S." was at Fifty-ninth Street, off Columbus Avenue, and was far from being the modern medical school of today. The old building itself was informal and our student life followed its pattern. I used to work there far into the night as well as Sundays and holidays. Two o'clock in the morning often found me bent over my dissecting table under a single drop light, surrounded by corpses in various stages of disrepair. Next to my table, a young man was working on the comparative anatomy of the nervous system. Dr. Huntington said to me one day: "You watch that fellow. He is the most brilliant student I've ever had in neurology. He'll be a great doctor some day."

The young man was the late Dr. Frederick Tilney. He fulfilled every prediction of his professors and became one of America's leading neurologists. Two or three times a week, Fred Tilney and I would be eating hamburgers and drinking beer in the gray dawn at a little restaurant near the college, for he, too, was a night owl.

In those days we sometimes got the corpses of murderers who had been electrocuted at Sing Sing prison. The electric chair had not been in general use for very long and the doctors were still interested in seeing what happened to the internal organs when such a high voltage went through the body. I used often to help the pro-sectors who prepared the corpses for dissection in the college morgue. I worked on some very notorious murderers but I did not keep any notes and for the life of me I can't remember their names. When the body was ready it would be sent upstairs to Professor Huntington. The graduate students would all gather round while he opened it up and showed us what the "innards" looked like in comparison with those of a normally defunct individual.

It was difficult for me to get used to the smell of dead human flesh. I would wash my hands in a strong deodorant but hours later the smell still seemed to be there. Because of that, I became a confirmed smoker. A cigarette or pipe always had to be in my mouth or I couldn't get through a long engagement with a corpse.

Comparative anatomy thrilled me. To take a muscle, for instance, in one of the lower animals and trace its development and modification through ascending groups right up to man was absolutely fascinating. It was my interest and enthusiasm, I suppose, that made Dr. Huntington and his scholarly assistant, Dr. Herman Schulte, urge me to go into surgery as a career. I might possibly have done so had not an expedition to the Dutch East Indies been presented just at that time. I know now that it would have been a fatal mistake for me to have become a surgeon. Psychologically, I was not fitted to endure the grind and confinement of a doctor's life. There were other fascinating bypaths that beckoned me away from my main job but I had sense enough to realize that they would lead nowhere in the end and that I must stay on the road I was born to follow.

I began to lecture publicly immediately after my first appearance before the New York Academy of Sciences. New York's Department of Education sponsored a winter series of free public lectures under a choleric supervisor named Dr. Leipziger. The fee was only ten dollars and often the expense of getting to the place left only seven or eight dollars net. I enrolled, however, for the winter of 1908, presenting my whaling pictures as bait. My first assignment was the Five Points Mission in lower New York. I knew nothing about Five Points or what one was expected to wear at those lectures, and went there in full evening dress, white tie, white waistcoat and all the rest. The audience gradually filtered in. Most of the men were coatless and without collars; the women were in corresponding negligee. I felt like nothing on earth in my tail coat. There was a long delay and I asked the "chairman" why he didn't begin.

"We are waiting for the police," he said. "Usually there is a riot if they don't like the lecturer. They throw things at him."

I got more and more uncomfortable and looked around to see what "things" they were likely to heave at me. Finally a policeman arrived and I was introduced to the audience. Something had to be done, I felt sure, for my shirt front would just invite a tomato or a cabbage. So I began by saying it was hot in the room and if they didn't mind I'd get myself more comfortable. I stripped off my coat, vest, collar, and tie, and opened my shirt in front. This put me more in harmony with my surroundings. Nobody said anything but there were a few smiles, particularly from the women. Then I cut all my preliminary talk and started full steam ahead with the pictures. It went all right, too. I tried to make them feel and hear the rush of the sea, the roar of the gun, and the thrill of the hunt. Never did I work harder, for there was the ever-present possibility of being plastered with overripe eggs if the story bored them.

The whole audience stuck it out to the end. Several men and women sidled up in an embarrassed way, while I was reclothing myself, and thanked me for the lecture. I felt it was a real triumph, but white tie and tails were out for the rest of the Department of Education lectures.

My attitude toward lecture audiences was changed very early by Mr. William Glass, who for many years was manager for J. B. Pond. He was a big hearty chap who used to say that he was the most lectured man in the United States, if not in the world. I guess he was, too, for it was his business to listen to at least one of the lecturers on his staff almost every evening.

I applied to the Pond Bureau. The first time Glass listened to my lecture he said it was excellent and they'd sign me up. Fortunately that night I had a good audience, one that reacted spontaneously and helped me along. The next time he heard me, a few weeks later, the audience was dead. I felt it the moment I started to speak. No one moved, or coughed, or blew his nose; they just sat there without the slightest reaction. My lecture was just as dead as the audience. Afterward I joined Glass.

"You certainly gave a rotten lecture tonight," he said. "In all my experience I can't recall ever having heard a worse one."

"I guess you're right," I replied. "But what could anyone do with an audience like that? Nothing I could say did any good. They died on me before I began."

"True enough; it was a bad audience, but that's no excuse. The committee paid for your lecture. They were entitled to the best you are capable of and you didn't give it to them. When a dead audience confronts you it means that you've got to work that much harder. If, at the end of every lecture, you can say, 'Well, I gave them the best I've got,' you have done an honest evening's work; if the audience doesn't appreciate your offering that's not your fault."

I hadn't thought of it that way and it made a deep impression on me. Since then I have never stood before an audience without trying to do my best. I don't mean to imply that all my lectures were good. Far from it. But when they were bad it wasn't through lack of personal effort.

In the spring of 1909, I made a short expedition to the St. Lawrence and Saguenay Rivers for the white whales. These beautiful porpoises come into the river in the spring, although they are a true ice species, and Dr. Charles H. Townsend, director of the New York Aquarium, wanted me to bring one alive to New York for exhibition in their great central pool.

My instructions were to get live porpoises, if possible. If not, I was to bring back skeletons, plaster casts, notes and measurements for a life-size model in the American Museum. I planned to suspend the live specimens on broad strips of canvas in narrow crates filled with water. The animals must not rest upon their sides or breasts for since they are accustomed to living in a supporting medium the weight of their own bodies would so press in the weak ribs that the lungs could not be properly inflated and they would suffocate. It seemed a practical project, for they are only twelve or fourteen feet long.

The animals were being taken in nets near Tadousac at the mouth of the Saguenay River for the sake of the oil and the hides, which the French Canadians sold for leather. Wind and weather conditions must be just right or the porpoises will not come into the tide rips where the net can be employed.

The conditions were not right when I arrived and the men seemed to think there was little probability of an early change. After waiting ten days I went off with three natives on their yawl to kill some porpoises for the Museum. We sailed along comfortably until we saw porpoises feeding in the tide rips. Many times we passed the brown young ones, but I did not want them. None but the old white fellows would do.

At last we discovered a school industriously catching fish at the end of an island. I put off in a canoe with one of the natives to paddle. My weapon was an old ten-gauge shotgun loaded with a lead ball. Some of the whales were young and we passed them, but a big white fellow slipped under very close and headed directly for us. He came up with a sharp "putt," as he blew and dived again. The next rising would bring him hardly twenty feet away. In a few seconds, the tell-tale patch of green water began to smooth out right in front. I fired the instant his round snowy head appeared above the surface. The beautiful animal shot into the air, falling back almost on the canoe. Dropping the gun, I grabbed a small harpoon and thrust with all my strength. At the touch of the iron the ghostly form again flashed into the air, but the native had tossed over the float and backed out of danger. The whale fought desperately to free itself, dashing from side to side and lashing the water into foam with its flukes. Watching my chance, I fired another ball into its neck. Straightening out, it rolled belly up and sank.

We got four other white whales that week. I took the skeletons and gave the men the skins and blubber; also a few dollars additional so that everyone was happy. The best specimen was beached well up in a sandy cove near Tadousac, where I could make plaster molds without interruption. The completed cast of my white whale now hangs in the Hall of Ocean Life in the American Museum. Besides the skeletons, I had enough new data on the habits of the species and photographs to make an interesting scientific paper.

Chapter 5

Yokohama's Yoshiwara

After returning from the St. Lawrence, I expected to settle down for a summer of hard work in the Museum. But, in less than a month, the director called me to his office. "Would you like to go to Borneo and the Dutch East Indies? " he asked. Would I? Would I like to go to Paradise?

For answer I almost leaped out of my chair. It was ridiculous to ask me if I wanted to go anywhere. I wanted to go *everywhere*. I would have started on a day's notice for the North Pole or the South, to the jungle or the desert. It made not the slightest difference to me.

"The U.S. Bureau of Fisheries have asked if I would lend you to them to go on a cruise of the exploring ship *Albatross*. They want to investigate the small islands of the East Indies and do deep-sea dredging. Your job will be to study the porpoises. Doubtless, there are many new species to be discovered. No one has done it. Also you are to collect land mammals and birds wherever possible. Better look up the ships at once. The northern route from Seattle will be the quickest."

It was all very matter-of-fact to him, but I went out walking in a dream. In the first place, the *Albatross* was the most famous ship of her kind afloat. No other exploring vessel was so well equipped for deep-sea dredging and her personnel had included some of America's most distinguished naturalists. To be numbered in that group was sufficient in itself even without the prospect of voyaging among the enchanted islands of the East Indies.

Getting ready in time to sail on the first ship took some doing. My personal arrangements were made in an hour, for I kept a small army trunk always packed and had a room by the week in the Sigma Chi fraternity house at Columbia University. But there were some items of field equipment which could be purchased only in New York. Nevertheless, when the S.S. *Aki Maru* of the Nippon Yusen Kaisha sailed from Seattle, I was on board. She wasn't much of a ship judged by those on which I crossed the Pacific many times in later years, but to my youthful eyes her luxury was superlative. I stood on the deck, watching the tide of passengers flow up the gangway. Japanese there were by the hundreds, bowing and sucking in their breath; a group of boys from the University of Wisconsin, the first American college baseball team ever to go to the Orient, and a sprinkling of lovely girls in bright summer dresses. But when the "all ashore that's going ashore" sounded, every attractive girl dashed for the gangway, leaving only a group of missionaries.

It made little difference anyway, as events proved, for the voyage was one of fog and storm, with one gale following close on the heels of another. Almost everyone was seasick every day, and I led all the rest. If I did struggle on deck the sight of those missionary women, green and disheveled, sent me below again with a new and worse attack.

When the *Aki Maru* drifted up beside the long wharf at Yokohama, hundreds of Japanese, chattering like monkeys, battled their way on board. Many of our passengers were Japanese businessmen returning from a long stay abroad. The greeting between them and their families was a marvel of restraint. Approaching each other, the bows and hissing would begin yards away. The wife had to stay down longest as she was of lesser importance than the husband, and it amused me beyond words to watch her peek to see if he was on the way up. Never a kiss or even a touch of the hand—just bows and smiles and hisses. I will say, however, that the tiny women and the children in their charming kimonos were just as attractive as their men were ugly. As long as a Japanese girl stays in her kimono she is a delight to the eye. Why the Japanese, with their inherent love of beauty, could ever have adopted foreign dress in their own country, I cannot understand. The kimono for both men and women is so graceful in itself and so comfortable that I should think they would have kept it in spite of their frenzy to westernize themselves.

Almost from the moment I set foot on shore at Yokohama, I felt that I belonged in the Orient. Even the smells, which most foreigners abhor, enticed me. There is a particular odor always associated in my mind with every country of the Far East. It may be that of the food that is cooked there, or of the people themselves, but if that particular scent were put in a bottle and presented to me anywhere in the world, I could tell you to what part of the Orient it belonged. We ourselves have just as characteristic an odor. The Chinese say we smell like sheep!

In Yokohama there was the Grand Hotel and "Number Nine." The former was the meeting place of the Orient and represented the nth degree of respectability; the latter was the most famous house of prostitution in the world and, in its way, was equally as important to the eddying currents of cosmopolitan life which flowed through its doors.

In the Grand Hotel was "Martin"; in "Number Nine" was "Mother Jesus."

Martin was a "runner" of mongrel ancestry (I think mostly Portuguese) but he had a fluent tongue and an uncanny understanding of the Japanese mind. When your ship was warped into the long open wooden pier which projected for hundreds of feet out into the harbor at Yokohama, Martin came aboard. He was immediately besieged by every foreigner on the vessel. Above the babble of the stevedores, and chattering Japanese, one heard the cries of "Martin! Martin!" How he ever kept his mind in such a flood of importunities is a marvel which only Martin could know. But when you finally captured him, like dragging out a dress from a bargain sale, Martin was always suave and unhurried. To him you surrendered the keys of your trunks, whispering in his ear the things you particularly did not want the Customs to see. Like the elephant, he never forgot. Your trunks arrived intact and were placed in your room in their virginal beauty. "Martin, of the Grand Hotel"—ah! He was the oasis in the desert, the draught of water to the thirsting traveler, the one known tie in a land of strange men and things. Afterward, in the long glassed-in veranda overlooking the bay with its myriad sails and ships flying the flags of every land, one had cocktails, or English tea. *"Boy-san, boy-san"* echoed through the lounge, but above all the rest was still the cry of "Martin."

Mother Jesus of Number Nine was just as important as Martin but in a different way and to fewer people. Number Nine was a house of prostitution, to be sure, but it was much more than that because of the woman who had made it famous. It stood at the far end of the Yoshiwara, the street of ill fame, in Yokohama. It was a broad street, lined with impressive houses, each fronted with a long narrow gallery and bamboo bars. After dark when the lights threw a yellow glow against the sky, these bamboo cages flamed with color. A row of lovely girls, each sitting behind an *hibachi* (charcoal stove) on the clean white mats, enticed the stranger or the young man about town. The color theme of one house was royal purple; of another peacock blue; still another rose, or pink or yellow. The passers-by looked and laughed and joked with the maidens sitting so demurely behind the bars. When a man was attracted by a girl, she came to the front of the cage and they talked with each other. If it went beyond the stage of conversation, he stepped inside the doorway where he made a bargain with the steward of the house.

The girl had nothing to do with financial matters, for she was as much an employee as any clerk in a department store. Her earnings went to the house, except for what gratuity she might be given. The patron was escorted to a room, furnished with a kimono, and was served the drinks or food he ordered. In due time, the girl of his choice appeared.

In those early days, the prices of each house that was keen for foreign trade were boldly advertised on huge street signs in English: "Short time, three yen; all night, five yen, including breakfast." A yen was fifty cents of American money.

The Yoshiwara was as much a tourist sight as the Diabutsu of Kamakura, or the Imperial Palace at Tokyo. There was nothing offensive about the street for police kept perfect order, and I don't remember ever having seen a drunken man. Japanese women seldom went there, but whenever a ship was in, the street was thronged with foreign girls, wide-eyed and thrilled at their close-up view of Oriental vice. Colored postcards of the little ladies in bamboo cages could be bought in any shop and ten times as many as of any other scene were sold to tourists.

But Number Nine was not like the others. Standing alone at the far end of the street, there were no bamboo cages in front of its dignified portals. A huge two-storied house, built around an open court, it boasted one of the loveliest gardens in all Japan. Its chrysanthemums were famous throughout the Empire and no better food could be had in Yokohama than was served at the tables on the terrace.

One had there what one wanted. It might be to dine quietly with another man where there was gaiety and laughter and the tinkling music of *samisens* (guitars). Perhaps it was to see a true geisha dance or to entertain a group of friends as one would do at any restaurant. If one wished to spend the night with a girl companion that was ten yen (including breakfast). But one was never importuned.

I went there on my second night in Japan with an Englishman who had known Mother Jesus ever since she was a child.

"You will come to the Orient often," he said. "It is a disease, you know, like malaria, only one seldom recovers. You have been infected. I can see it in your face. Better, then, that you should know Mother Jesus at once. She will be useful to you as she is to all her friends."

So to Number Nine we went and ordered dinner of *suki yaki* in a beautiful little room opening off the gallery above the garden.

He sent his card to Mother Jesus and when our meal was ended, when by no possible effort could we have eaten more, a screen slid back and our hostess came. I don't know what I expected her to be like, but certainly not what she was. She was barely thirty, slim and graceful, not beautiful, or even pretty, but strangely attractive with calm appraising eyes behind which seemed to lie the wisdom of the ages. I was introduced and we talked of my voyage across the Pacific, of this and that. Our conversation was such as might have taken place at any dinner table on Fifth Avenue for her English was as good as mine. When we rose to go my friend said, "Fujiyama showed her head above the clouds when he sailed up the bay yesterday, so, of course, he will come to Japan again. I hope you will be his friend and help him if you can."

In this way I first met Mother Jesus. Her special friends were the captains and officers of the foreign coastal ships and ocean liners with a sprinkling of world travelers such as I became. For them she was a clearing house of information. She delivered private messages, kept their money, paid bills or debts, and gave them good advice which probably they never took. Whenever a ship came in, Mother Jesus knew just who its officers were, heard all the gossip, and could arrange a passage even when the boat was crowded. These things and many others she did for those she liked. In return, they brought her presents from every corner of the world and spent money in her house as only sailors will.

During the fourteen years of our acquaintance, we became good friends. I know she liked me, and I not only liked but respected her. But there was an aloofness and inscrutability about her that was strangely baffling. One couldn't tell what she really thought or felt. Never did she lose her quiet dignity, even under the most trying circumstances. She maintained, too, a certain dignity in her house.

I asked once about her past, but she would tell me little. Only that she was born in Number Nine thirty odd years before, had never married, and never had a lover. Strangely enough, I believed her, though it seemed incredible.

The Yoshiwara which tourists knew gradually disappeared as Japan, more and more, adopted Western customs. At first shutters were put up on the street side of the bamboo cages and visitors could see the girls who still sat behind their *hibachis* only from inside the covered entrance to the house. Then even that was abandoned and an applicant was ushered into a small room where he sat at a table smoking while the girls trooped in, eight or ten at a time, and lined up against the wall for his inspection. A most unattractive performance it was, at any time, and particularly so after the brilliant display of earlier years, when there was something of romance about the Yoshiwara.

In the old days, girls often found husbands of no mean estate there in the Street of Joy. Probably few of the girls went into the houses of their own volition. Many came from the families of respectable farmers or small tradesmen who had lost their money or got themselves involved in debt. If they had an attractive daughter she could be sent to the Yoshiwara on a "lend-lease" basis for a period of years. Sometimes it was for three, sometimes for five years, depending upon the sum of money needed by her family. If by any chance their fortunes had improved sufficiently before her lease expired, they could buy her back, but at a considerable profit to the owner of the house. If a client fell in love with a girl he also could buy her contract and this happened not infrequently. I was told that if a man of fortune married a girl from the Yoshiwara he lost face somewhat but that eventually the origin of his wife might be forgotten and she might be accepted by the social world.

Of course a geisha was, and is, quite different from the ordinary prostitute. Geishas correspond exactly to our night-club entertainers. They are girls who, when very young, show a certain flair for singing or dancing and study to improve their talent. Also they are taught all the social arts, and if their wits are quick, they develop such repartee and small talk that they may gain national reputations. At dinners they may say things with impunity which no one else would dare, often taunting important personages with knowledge of their intimate private affairs. A geisha may be chaste or have a lover, or be promiscuous—it is for her to choose.

The first big Japanese dinner I ever attended was given in my honor at Shimonoseki by the British Consul. He had lived there for many years, spoke Japanese like a native, and after he retired, died in his lovely house on the hill overlooking the Inland Sea. He loved entertaining and he did it extraordinarily well.

I was his excuse for a real Number One party which he gave in the biggest Japanese hotel. He gathered all the gay blades of the town (foreigners) of which there were only about twenty in those days. When we arrived, each was given a beautiful kimono and shed his pants, for there is no more uncomfortable garment in which to sit upon the floor than a pair of trousers. They were checked with our shoes at the door. Then we repaired to the banquet hall, a huge rectangular room. It had been arranged merely by removing a lot of sliding screens which ordinarily divided the top of the house into separate cubby holes we would call bedrooms. There was a little table for each of us arranged in a semicircle and a whole flock of Madame Butterflies flitting about laughing and chattering. My particular little insect alighted in front of me unerringly and announced in broken English, "I am you."

"Well," said I, "that's fine. How did you find me so quickly among all these foreigners?"

"Oh," she giggled, "that was easy. I have the honorable photograph." With that she extracted from the sleeve of one of her kimonos the most God-awful newspaper picture I have ever seen. It was from the *Tokyo Shimbun*. Even my own mother would not have recognized it. Then "Cherry Blossom," for that was her stage name, proceeded to recite to me all the details of my past life which the newspaper had published.

This took place after we were seated on the floor, each behind his little table with his own particular *musume* to wait on him. She was his for the night and apparently she considered it an honor to do his slightest wish. Steaming hot *saki* was poured into tiny cups from little whistling jugs, and then came the first course of prawns—great shrimps rolled in egg and cooked in deep fat, than which, I don't mind saying, there is nothing more delectable even if it is Japanese. The dinner went on with more and more *saki* and delicious chicken *suki yaki*.

Half a dozen geishas came in and did a song and dance which left me cold—a nasal whining which no foreigner can appreciate and a series of postures supposed to portray an iris, a lily, or some other flower swaying in the wind, by movements of the hands and graceful sweeps of the kimono. It was all very artistic and for a dilettante of the arts would, I am sure, have produced ecstatic thrills. Anyway, by that time we had had so much *saki* that we applauded it properly and the geishas retired well content.

I noticed every so often that some one of the men got laboriously to his feet and disappeared into the main hall. Always his particular little shadow accompanied him. I supposed I knew what he was about but I couldn't understand the girl companion. Eventually I said to her, "I gotta go. I gotta go." Apparently she understood for she said *"Hai"* (yes) and taking me by the arm conducted me to the *benjo* which was open and distressingly public. She got a basin of water and a towel and stood there expectantly. I tried to shoo her away but she wouldn't be shooed. By that time I'd had enough *saki* so that I guess my inhibitions were somewhat anesthetized. I discovered that the basin of water and the towel which she was so patiently holding were for what you pay ten cents to the attendant in the hotel washroom when he fills the bowl and brushes your clothes. She wasn't a bit embarrassed; it was just custom. It was good training, however, for in later years I was able to argue brazenly with a lady in a Paris theater anteroom, bound on a similar mission, as to who was first. After all it is simply a matter of physiology.

The dinner didn't end until a late hour. I said good-by to Cherry Blossom (I thought). She recovered my pants and shoes and helped me put them on. Also she established me in a rickshaw. Then someone suggested that we stop at the Shimonoseki Club for a nightcap. It was an hour or more before I reached my room in the Sanyo Hotel and the lights were on. There sitting on the bed was Cherry Blossom.

"What are you doing here?" I asked.

"Why, Griffith-San engaged me for the night—for you."

"The devil he did! I'm sorry but I don't want anybody. I'm drunk and I'm tired and I want to go to bed—alone!"

Then she began to cry. That was terrible. I'm a pushover for a woman that cries. So we talked and I discovered that Griffith-San had paid for all the girls to spend the night with all the men. If she went back she'd lose face because her employer would think she hadn't made good with me, the guest of honor, and oh, it would be terrible, terrible! As I said, I'm no match for a weeping woman.

"There's the other bed," said I. "Suppose you just get in there and don't bother me. All I want is an aspirin and sleep. Good night."

From the moment I touched the pillow I don't remember another thing. It was noon when I opened my eyes, and Cherry Blossom had gone. I never saw her again.

Chapter 6

Robinson Crusoe's Isle

When the *Aki Maru* left Japan and nosed her way southward through the Inland Sea, she stopped briefly at Shanghai where I got my first sight of the country I was to call home for nearly a score of years, though I didn't know it at the time. It wasn't a pretty view coming up the river to Shanghai, but the water life was fascinating. Every junk and sampan has an eye painted on either side of the bow. Why? "No have eye, no can see" is the logical Chinese answer.

I was amused, too, at the frantic maneuverings of the sampans to avoid crossing the stern of the *Aki Maru*. The boatmen, it seems, believe that a not very nice devil dogs the footsteps of every person and if there are five hundred people on board, five hundred devils follow the ship. At all costs, therefore, a sampan must keep from crossing the immediate wake of an ocean liner; otherwise, the horde of accompanying devils will be switched over to his boat. If the worst happens and he does cross, he must shoot athwart the bows of another sampan at once to divert the evil spirits to the other fellow. It develops into a regular game of tag which is all the more amusing because it is so deadly serious to the participants, who scream and curse like mad while jockeying for position.

I didn't like Shanghai at my first visit, and liked it no better on more intimate acquaintance. I don't know what it has become since the war, but then it was a big cosmopolitan city sprawling over a mud flat with a veneer of second-class England and Europe on a background of Chinese civilization. Its foreign life was, of course, dominated by the British. The American Club and the French Club were both nice places where one met one's own nationals, but the Shanghai Club dominated the social life of downtown. It boasted the longest bar in the world and I guess it was. If one wasn't a member of the Shanghai Club one didn't quite "belong." The city's social distinctions were so complex that it was difficult for an American to figure them out.

For my sins I had to land there often on trips to and from America and sometimes came down to play polo, but I always left as soon as possible. I don't think I ever spent a really happy day in Shanghai, for no matter how nice people were, they were not able to create a really friendly background.

The *Aki Maru* went on southward to Hong Kong which I suppose now is a wreck of its former self. In many trips around the world I have never seen a more beautiful harbor, or one which gave such an interesting picture of cosmopolitan shipping, for I counted the flags of almost every maritime nation on the earth when we dropped anchor in the bay. Contrary to Shanghai, I liked Hong Kong instantly, although, God knows, it had an official community that was stuffy enough. But that was only a small part of it and I always have been happy there. At the club I met a charming Englishman who was a passionate amateur naturalist. Without the slightest embarrassment, he said, "I'd love to ask you to my house for I have a wonderful collection of live pheasants, but I don't know how you'd feel about it. You see I'm not married to the lady I'm living with. Devoted to her and all that, but she's half Chinese. Father a Scotchman. Just couldn't marry her, you know."

I wasn't embarrassed by his frankness either and assured him that I didn't give a tinker's damn about his domestic affairs but I would like to see his pheasants. To myself I had to admit that I was keen to learn how a ménage without benefit of clergy worked in the Orient.

We went over to Kowloon and out to a charming bungalow on the hillside. Behind the house were twenty or more wire pens with the most superb collection of pheasants I have seen anywhere in the world. Just before tiffin, gin and bitters were served on the wide veranda and there I met the lady. She was perfectly lovely with blue eyes, a skin whiter than mine, and a divine figure. No one on earth would have suspected mixed blood. Her childhood had been spent in Scotland, at school, and she had returned to Hong Kong to be with her father, a retired sea captain who lived a mile away. After tiffin, however, I saw the other side of the picture, for her elder sister walked over with a fourteen-year-old brother. Both of them were very dark with slanted brown eyes and it was difficult to believe that they were not pure Chinese.

Next day I learned more about my Englishman from the American Consul. He was a well-known exchange broker and a highly respected member of the community. He would remain so just as long as he did not marry the girl. His social life was entirely separate from that of his common law wife and he was treated as a bachelor by the foreign residents of Hong Kong. Five years later I returned to find that he had died of typhus. Of his beautiful mistress I never heard.

Such was my first contact with what will always be one of the greatest social problems of the Orient. It is a most distressing problem, too, because from the standpoint of eugenics such race mixture has not produced good results. I was told that as a rule half caste children were more than usually brilliant up to about fourteen years of age. Then their mental development seemed to slow down, or to almost stop, and it was seldom that in later years they had a record of achievement. Sometimes the half caste girls were very beautiful, challenging any man to keep from falling violently in love. But marriage meant social ostracism from both races and eventual unhappiness. Children were unpredictable. They might resemble either parent or be a mixture of both. It was all a gamble.

Leaving the *Aki Maru* at Hong Kong for the S. S. *Tamin* we rolled across the China Sea to Manila in the wake of a typhoon.

These circular storms hatch off the Philippines, sweep over toward Hong Kong and up the China coast often as far as Japan. I've been through three—two at sea and one ashore. There is something majestic but utterly terrifying about them. Wind of a force beyond belief and torrents of rain. It is an experience to have once—but only once.

At Manila I seemed transplanted into another world—that of Spain. The modern city of today was just coming into being. The Army and Navy Club occupied a picturesque Spanish house in the walled city; the moat was still a slime-filled ditch. I rode up the Escolta, Manila's main street, to the old Metropole Hotel on the Santa Cruz bridge in a *carametta* drawn by a diminutive but very active pony. There I shaved and had breakfast, the first food that had stayed put since we left Hong Kong three days earlier.

As a companion there was Lieutenant Treadway of the Philippine Scouts, a tall raw-boned Texan whose stories of life in Moroland I had listened to with fascinated horror between bouts of seasickness as we plunged across the China Sea. I remember particularly one tale of a buddy he had roomed with for three years. The young officer had walked beyond the limits of the fort one evening when all was apparently peaceful. He didn't return and next morning Treadway was leader of a searching party that found him buried in an ant hill up to his neck. His eyelids had been cut off, his tongue pulled out by the roots, and a train of honey led to his open mouth. There wasn't, said Treadway, anything left except a horrible reminder of what had been a man. For two days and nights they tracked the murderers and made them prisoners, but as he said, "All ten were killed while trying to escape."

He told, too, of a Moro who had run amuck in the main street of Jolo. Treadway heard the screams of people who were being cut down by the fanatic native and rushed out just in time to meet him leaping down the street waving his bloody *kris*.

"I began shooting with my .38 Colt revolver when he was thirty feet away," Treadway said. "Every bullet went 'plunk' into his body, but he kept coming. I had only one left and he was hardly six feet from the end of my gun. That caught him squarely in the forehead and lifted the top of his skull right off. He dropped with his *kris* almost touching me."

In 1909 the "Big Man" of the Philippines was Dean C. Wooster, secretary of the interior. He exercised more actual power than the governor himself. Wooster was a well-known ornithologist and during early bird-collecting expeditions had traveled over most of the islands and knew the natives as no other official ever did. He was a forthright man who left no doubt in your mind as to where he stood.

The *Albatross* was in the south near Zamboanga and wouldn't return for several weeks. I wanted to get busy at once so Wooster told me of a small island, off the track of coastal vessels which he had long wanted to explore. He would, he said, arrange to have a government steamer drop me there and pick me up on its return trip.

A week later, on a glorious tropical morning, I was rowed with two Filipino boys toward the low shores of a palm-clothed island. Water, green as emerald, covered the outlying coral reef over which floated fish painted in rainbow colors. We landed on a sandy beach in a little bay and made a rapid reconnaissance of the island. It was uninhabited but had a spring of good water; that was the important thing. After leaving enough food for five days and our collecting gear, the ship steamed away.

Again I felt all of the sensations of wonder and joy I had known when, as a child, I had listened to my mother reading *Robinson Crusoe* to me over and over again, and I, in fantasy, had lived on just such an island as this. Now here I was, with my childhood dreams come true. Only instead of one, I had two "men Fridays."

We made our camp beside a huge rock and swung ship's hammocks from the branches of an overhanging tree to be well away from the land crabs. Disgusting creatures, these giant crabs, which swarm over a wounded animal, literally eating it alive, or one that is dead, if they are a little late in getting there. Half the specimens caught in my traps during the night were devoured before I could rescue them in the early morning. There was only one family of monkeys on the island and a few smaller mammals, but the place was alive with birds. Parrots flashed among the trees and beautiful cream-white pigeons with black wings and tails filled the air with soft cooings and fluttering wings.

Each morning I was up at the first rosy flush of dawn to run the traps, explore every nook and cranny of the island, and shoot new birds. Then back to camp for a swim off the beach before settling down with the boys to skinning and preparing the day's specimens. In the afternoon we waded the tide pools, collecting fish, crabs, snails, and everything that moved and was alive. The glorious weather held without a break. It was hot during the day, of course, but we wore nothing but a pair of trunks, and the nights were always cool.

For five days I lived in a glow of ecstatic happiness and bemoaned the fact that the ship would return to end my island dream. But on the fifth day the sun set and the velvet darkness came down about us like a curtain, and no vessel appeared. The next day came and went, and the next. Still no ship. Our food was gone—even to the last ship's biscuit—but I still had a few shotgun shells and those gave us pigeons.

The Filipino boys began to worry. I didn't care; I was too happy. When the ammunition was gone, I had the natives weave a great net out of palm fibers. This we strung over a favorite roosting tree of the black and white pigeons and the first evening snared more than fifty. Fish were easy enough to catch on the reef and we evaporated sea water and found salt along the edges of the tide pools. With pigeons, fish, crabs, and salt we certainly could not starve.

For two weeks we lived like castaways. Then one day a streamer of smoke showed against the sky and the tiny vessel nosed her way toward the island. With my glasses, I could see the captain pacing restlessly back and forth on the bridge, but we stayed hidden. A boat dropped over the side and two sailors beached it in our little cove. They found us just dipping into a pot of pigeons, stewed with palm roots, and we gave them a share served in half cocoanut shells. Packing up our few belongings at the camp beside the rock wrung my heart, for I knew that never again would I have such utter content as had been my lot that wonderful fortnight. The captain I found in a fever of anxiety. When he learned how happy and comfortable we had been he was utterly disgusted. The ship had been delayed by a damaged propeller and he had worried frightfully for fear we would starve on our desert island.

Eventually, I joined the *Albatross* at Cavite Navy Yard. She was a beautiful ship, built like a yacht, with a wide afterdeck where the officers slept on camp beds when the night was hot. It seemed almost a dream when I awoke the first morning in the brilliant flush of a tropic dawn to hear the boatswains' silver whistles piping the men to quarters on half a dozen warships riding at anchor a few fathoms away. The admiral's flagship, the *Rainbow,* lay just off our port quarter. One of his lieutenants, Whiting, a marvelous swimmer, had set the whole Navy agog just the day before my arrival. He had demonstrated how, in a crippled submarine, all hands could be saved except the last man. Crawling into a torpedo tube, he ordered the hatch shut behind him and then opened the forward door. After the compartment had filled with water, he crawled out of the tube and rose to the surface. Whiting, only that morning, had received the congratulations of the Secretary of the Navy.

I was invited to a big dinner in his honor given by his brother officers on the *Rainbow.* It was a frightful night and even in the shelter of Manila Bay the waves were dashing against the anchored ships in a smother of foam. The evening was very wet indeed, inside as well as out, and Whiting was on the "crest of the wave." While we were having coffee, he decided it was just the moment for a swim. Without saying a word he ran topside and plunged overboard in full dress uniform. The ship was in turmoil for it seemed that no human being could live in that tremendous sea. A dozen life belts were thrown over, but Whiting bobbed up alongside and climbed aboard happy as a duck. He couldn't understand what all the excitement was about.

We moved out the next day, and I did not see him again until twenty-five years later at a dinner of the Camp Fire Club, I think it was, in New York City. That evening I sat next to him while he related the story of his submarine exploit in 1909. After his very modest recital, I told the Club of his midnight swim in Manila Bay, much to his chagrin.

Life on the *Albatross* was new to me and interesting. She was a "bastard" ship, according to the Navy, for she was owned and controlled by the U.S. Bureau of Fisheries but manned by the Navy. None of the officers cared for the duty because they felt that, while it was pleasant enough, it did not advance them in their profession. It wasn't a "happy ship." Most of the scientific staff as well as the officers had been aboard her too long, and friction had developed to such an extent that several were not on speaking terms with the others.

By a lucky chance, I got off on the right foot with both officers and men. The executive officer was a lieutenant who had been a famous pitcher on the Naval Academy's baseball team. The *Albatross* had eight good ball players but no one who could catch Barthlow. Although my position in college baseball had been first base, I volunteered to try as catcher and in practice did all right. Barthlow slammed them in like bullets but somehow I managed to stand up against the barrage. As a result, the *Albatross* challenged any other ship in the Navy Yard to a match. The *Rainbow* accepted and the game took place on a Sunday just after pay day. A lot of money had been bet on us by the *Albatross* crew and they cleaned up. We won six to two. Since I had supplied the missing link in the team, it went a long way with the men. That was the last time I ever played serious baseball.

Chapter 7

Typhoon

The *Albatross* steamed southward from Manila, through the Sulu Sea with only a momentary stop at Twao, British North Borneo, and then on to Sibattik Island for coal. It was a breathless day of torrid heat when she dropped anchor in Sibattik Bay. I can see in memory the sheer wall of jungle; the giant camphor wood and "king trees" stretching up and up till their summits seem to touch the sky; the palms and creepers and ropes of vines, bristling with thorns like the barbed-wire entanglements of a battlefront. I can hear the myriad singing insects which filled the air with such a medley of shrill vibrations that my ear drums ached. Then at four o'clock in the afternoon all was still. Utter silence lay over the jungle while great billowing clouds rolled swiftly in from the sea, and torrents of rain poured out of the sky. Abruptly, the deluge ceased, to leave the jungle steaming like a caldron. Then the insects began again, shrilling louder than before, and the forest creatures took up their separate lives where they were interrupted by the flood of rain.

I remember, too, my first attempt at moving through the jungle. I tried to force my way—but once only. Thorns and barbs caught and held me in a dozen places; "wait-a-bit" vines laced my chest and dug deep into my arms and legs. Every move was agony. There was nothing to do but detach them one by one and cut my way to freedom.

A surprise of the present war is how armies have been able to fight in the jungles of the East. I remember talking years ago to British officers about the defense of Singapore. To the land side they merely waved their hands and smiled.

"There is the jungle! It's better than any man-made fortification; no troops could move through that."

It seemed so at the time, and yet the Japs did it. They slithered through the tangle of thorns and creepers with the lizards and pythons, like some loathsome half-human creatures born of the primordial slime, each one cutting his own path, worming his way behind the British lines. The jungle didn't stop them, nor the fever, nor the leeches. Swarming on the leaves, in the bushes and on the ground, the blood-suckers work their way through every tear or hole in your clothing, even through the eyelets of your shoes. Moreover, the wretched creature deposits a serum which prevents the blood from coagulating. The wound stays open and is certain to become infected; then you have a nice mess.

All my life I have loathed snakes; I have to force myself to touch them and yet I have injected and brought back hundreds for the Museum. Of course, I supposed there would be dozens of snakes in the jungle and dutifully I did try to find them, but the vegetation was so thick and many of them are so protectively colored that they eluded me completely. There was, however, one shining exception. I was following a deer trail with my Filipino boy, Miranda, when suddenly he jerked me violently backward.

"Excuse, Master, but right there, big snake. You shoot him. Quick."

He pointed and I looked, but to no avail. I couldn't see any snake. A great tree overhung the path and he kept saying "There, there, don't you see him? Right on that branch."

Suddenly a breath of wind shifted the leaves and a patch of sunlight filtered down to rest upon a glittering eye, in a great flat head pressed upon the branch. Following it back I made out yards and yards of snake stretched along the tree trunk, poised to drop upon anything that moved. I backed off and lined my sights upon that shining eye. At the crack of the rifle, a typhoon seemed to have struck the jungle. A writhing, twisting mass of flesh and muscle mowed down bushes, crushed small trees, and wrapped itself in a mass of thorn vines and creepers. Miranda and I ran. At a safe distance, we watched until the storm quieted, then cautiously ventured back. The snake lay there, still jerking spasmodically, looped and coiled upon the ground. The head was smashed to a pulp, and when we straightened the reptile out I paced the length. It measured twenty feet. The stomach was empty and the great serpent must have been very hungry. It was lying above the game trail ready to throw its coils about anything that passed below. Without doubt, Miranda's sharp eyes had saved me from a horrible death.

From the first gray light of dawn until far into the night, I was busy. The ship's doctor advised me to rest as others did during the fierce heat of midday. But, in the enthusiasm of youth, I laughed at his warning until one afternoon when the temperature was one hundred and twelve degrees in the shade. Then, suddenly, I dropped under a palm tree in the steaming jungle. Black patches darted before my eyes and I was violently ill. It was only a heat stroke, caused by too much exertion in the heat and too little sleep, but it gave me pause. From then on, I did my shore-work in the early morning and late afternoon, spending the broiling hours of midday in the laboratory preparing specimens.

Because the *Albatross* had explored the sea bottom from the Philippines to Borneo, she did no hydrographic work until we crossed the Celebes Sea. In those waters, blue as indigo, she dropped her nets sometimes a mile, or even two, straight down to the ocean floor. Usually the dredge contained a great mass of ice-cold mud, but as this was washed away strange sea creatures began to appear from out the muck. Animals from the complete darkness of a submarine life where the water pressure was enormous were dragged to an upper world for which they were not adapted. There were fish with eyes far out on stalks; others bearing phosphorescent spots along the sides like the glowing portholes of a lighted ship;

fish carrying little lanterns in front of their noses to light the way. Sometimes in the sudden ascent to the surface and release from the terrific pressure they were turned almost inside out. Usually those from the greatest depths were badly damaged.

My job, of course, was land collecting and I was supposed to have no part in the dredging operations. But I could not stay away from the forward deck when the net came up from those mysterious depths. Always something appeared which set my imagination aflame. One day there was the figurehead of an ancient ship from the deepest part of this romantic sea. Made from the trunk of a great tree, it was fashioned in the likeness of a woman with streaming hair. So covered was it with barnacles that at first we thought it merely a water-soaked log and of no account. But I seemed to catch the shape of something human under the crust of sea life and with a chisel and a hammer cut away the barnacles that had bit deep into the solid oak. Gradually there emerged a woman's form, her outstretched hands clasped before the full breasts which still retained the shape of youth. How old it was or from what ship it came, there was no way of telling, but assuredly it was not of recent time. For a hundred years, or for centuries perhaps, it had rested on the ocean floor until our net dragged it from the eternal deep. What, I wondered, was its history? Had it graced the bow of a Spanish ship sailing in quest of the riches of the Indies? Or was it English, or Portuguese, or Dutch? Had the ship been lost by storm or sunk in battle?

For a week I kept the lady we had rescued from the sea. I wanted to take her home with me, to learn more of her history, if I could, from those who knew the story of ships and figureheads, but she was too big to save. I could not take her to my cabin for there would have been no place for me, and the quartermaster complained of her presence on the forward deck. To him she was just a log of wood and in the way. So I nailed upon her body, just beneath the breasts, a lead plate with my name, address, and date, and a plea that if she were found again I would be informed. Then from the afterdeck I consigned her to a new grave in the Molucca Sea.

Perhaps now she is resting beside the bones of our own American men and ships, for it was there that a great battle was fought with the Japanese in 1942.

As I think back to those lovely tropic islands bright with flowers and graceful palms floating like giant lilies on a sea of incredible blue, in peace unutterable, it seems beyond belief that only a few short months ago they could have echoed to the crash of cannon and the screams of dying men.

Menado, our first port after Borneo, was pure enchantment, a town made for exhibition, it seemed, not for use. It rests sleepily on the long arm which the Island of Celebes stretches northward toward the Molucca Passage. The harbor is only a semicircular bite out of the land and we were told that at times a tremendous swell rushes in from the Celebes Sea. But when the *Albatross* dropped anchor just at dawn, the water was as smooth as glass, mirroring a shore line of feather-like bamboos and lacy palms. The volcano behind the town stood out sharply in silhouette against a sky of vivid orange-yellow shot through with streaks of crimson. At the end of the long wharf, we stepped into a picture-book village. Wide streets lined with magnificent trees, perfect little houses, velvet lawns, and gorgeous beds of flowers. Roads which seemed to have been newly swept. Natives in brilliant sarongs and scarlet fez. Everything scrupulously neat and clean; "spotless town" if there ever was one.

It was a happy place. I sensed it from the natives who gave me a brilliant smile and courteous greeting to the pleasant-looking Dutchmen on their way to offices shortly after sunrise. In mid-morning, they retired to the coolness of shaded verandas and darkened bedrooms, there to sleep away the hours of heat. The streets were deserted save for a few brown-skinned natives. About five o'clock in the afternoon the white population bestirred itself. Women in filmy dresses and men in white duck clothes gathered at the club or drove about the streets until the soft darkness of the tropic night dropped like a stage curtain, shutting out the palms and flowers. But it didn't send them off to bed. Night was the time in which to live; the day for sleep.

We saw a dozen towns throughout the Indies all similar to Menado. Amboina, Gorontolo, Macassar, Ternate, Gillolo—they all gave the same picture of beauty, of cleanliness and of a happy people drowsing away the hours while the peaceful days went drifting by.

During the past thirty years, I have talked often of these little towns, but to everyone they sounded vague and far away. Never did I find a soul who had walked their flower-bordered streets and few who even knew their names. Then, less than half a year ago, I waked one morning in New York to see the word Menado blazing across the front of every paper. My tiny village at the end of the earth suddenly had become of world importance. Only the day before, Japanese planes had rained death and horror upon its quiet beauty and its gentle people. Out of the skies chattering, blood-thirsty little men, clad in khaki and steel, had dropped into its shaded streets and crawled about its gardens like a swarm of loathsome maggots.

Menado's name had hardly left the front pages of the world's press before that of Amboina took its place. The *Albatross* dropped anchor there on December 4, 1909, and today I read in my journal how after we had landed at the long wharf, I went up a deep canyon collecting birds and small forest animals. A flock of hornbills flew overhead "making a noise exactly like airplanes" I recorded. I shot a huge lizard lying on a branch over a deep pool from which I collected several fish of a new genus. And then in the late afternoon, it says that I climbed to the top of a hill where the bay and town lay spread out before me "like an aerial photograph."

While I was sitting there, loving the view and the trees and flowers, a caravan of ants crawled over my legs. I'd never seen any ants like those and I cracked one between my teeth for that is a good way to identify them quickly. The sharp acid taste was distinctive, so with a pair of forceps from my collecting bag I picked some off my legs and put them into a small bottle of alcohol. They went to Professor William Morton Wheeler at Harvard University. It represented, he said, an unknown species and he named it in my honor. I hope my ant helps to make life miserable for the Japanese who are now using Amboina as an air base from which planes roar out on errands of destruction over the Moluccas and the Banda Sea.

On the wild mountainous island of Buru, then but partially explored, we met the only unfriendliness from natives. The Dutch governor at Ternate had warned us about Buru, but when we landed, there was not a single native visible. I went far inland with two sailors along the edge of a stream bed. At several places we found Malay huts, evidently hurriedly abandoned, for fires were still burning and food half eaten. The strange feeling that unseen eyes were peering from the jungle made us definitely uncomfortable but never could we catch sight of a human being. We had been following a trail along the stream and when it was time to return I explored it cautiously. Sure enough, we found just what I feared. Sharpened bamboo stakes, probably poisoned, set at an angle along the trail, so they would jab us in the thighs. It was a Malay trick of which I had often heard. Abandoning the path, we worked down the stream bed to the shore. I, for one, was devoutly thankful to see the boat.

Every day of the cruise was filled with interest and excitement. Often I was dropped off on tiny uninhabited islands to spend a day or two while the ship dredged in the vicinity. There were new birds and animals of which I had read but never seen. We hunted the great sambur deer, as big as an American wapiti, the strange wild boar, babirusa, with tusks growing straight up through the snout, and monkeys of half a dozen species.

Christmas Day, 1909, found us in Macassar on the southern arm of Celebes. Never will I forget that day! The "Battle of Macassar," as it was known ever after in the Ward Room. The governor invited the captain, the doctor, and myself to go crocodile hunting with him in the morning. We divided up into three parties. I was alone and so was the doctor. I didn't get any crocodiles. I couldn't have hit one to save my life, but the doctor killed a big croc by mistake. His native paddler saw it asleep on a high bank.

"You shoot 'em in the head," he said. Dr. Lee tried to follow directions, but he wasn't at his best that morning. The bullet struck the croc in the tail and he made a wild leap off the bank, landing on the outrigger of the canoe. Mouth wide open he kept coming. The doctor stuck his rifle right between those gaping jaws and pulled the trigger. The back of the croc's head suddenly disappeared and the beast sank down half across the canoe.

When the *Albatross* returned to the Philippines and dropped anchor at the Cavite Navy Yard, all of us were wild to go ashore. For months we had been out of circulation and we wanted to celebrate in a big way. Every shop looked enticing, every girl was beautiful, and the music intoxicating.

The *Albatross* tarried less than a fortnight at Cavite because her three-year Oriental cruise was ended. As she sailed out of the harbor, a hundred-foot homeward-bound pennant streamed from the masthead. Every ship bade us God-speed with siren whistles, each officer tossed his cap overboard, and the blue-jackets set afloat a dummy on a raft bearing a placard "Good-by Manila." They fondly hoped, I was informed, that when it floated out to sea some ship would pick it up thinking that it was a stranded sailor.

Steaming northward, dredging on the way, the *Albatross* touched briefly at the southern tip of Formosa where I collected a few new birds and mammals. The little village of Soo Wan in the north was a strange place, half Chinese, half Malay, built of cobblestones right on the water's edge. Camphor was its mainstay in life. We could smell the odor far out at sea on the shore wind.

Soo Wan came near to being the last earthly port of call for the *Albatross* and all her crew, because we ran out of the bay full into a typhoon sweeping up the Formosa channel. It came with amazing suddenness, catching us when we were halfway to Keelung, forty miles from Soo Wan. In the beginning, we passed a small, low-lying British gunboat. She was only a few hundred fathoms off our port beam when I saw a man start aft presumably to hoist her colors. I was watching through my binoculars and saw a great green sea lap over her stern, sweeping the man off the deck like a straw. It was pretty awful, seeing that man disappear so quickly into the smother of white-topped waves! Neither did it help our peace of mind for we had begun to realize that the gallant old *Albatross* might end her twenty-eight years of service at the bottom of the sea right there. In a few hours, perhaps we ourselves would feel the strangling water closing over our heads. A tense calmness pervaded the ship. We were fighting a battle for life against the elements, and the odds were on their side. No one talked much for it was difficult to hear above the shrieking gale and crashing water.

The *Albatross* was headed directly into the seas which broke over the bow and swept the deck every time the ship dived into one of the mountainous green waves. A mile away, sheer cliffs rose like a wall above a narrow beach, smothered in white foam. For some reason, the captain had decided to fight his way against the rising storm instead of riding it out in the open sea. Keelung was only twenty miles away, but often we barely held our own. Foot by foot, the old ship crept forward, sometimes losing more than she gained, but always coming back for another assault upon the crushing waves. There was something distinctly personal about the fight. It was man against nature. Everyone on the ship was a part of the battle. I don't think I was frightened; no one seemed to be. All our minds and hearts and strength went out to help the *Albatross* when she staggered drunkenly after a smashing blow in the face.

Just as night closed in, lights showed on our port bow. An hour later the ship limped through the narrow entrance of the outer harbor, battered and bruised, but game to the last. Outside the typhoon roared past, increasing in violence every minute. Next morning, during the half-hour run to the inner anchorage, the starboard engine gave away. Had that happened before we reached shelter, nothing could have saved the ship. It was because the captain knew of the weakness of that particular engine that he dared not ride out the storm on the open sea.

It required a week to make repairs in Keelung; then we went northward to the Loo Choo Islands, that "Forgotten Kingdom of the East." I photographed the ancient Shuri Palace, in Naha, where Commodore Perry made a treaty with the King of Loo Choo when he "opened" Japan in 1854. The finest red lacquer ware of the Orient is made in Naha. Some of the trays and bowls I bought there we are still using at Pondwood Farm after thirty-two years. At Nagasaki, Japan, the voyage on the *Albatross* ended for me and a new and wholly different one began on land.

Chapter 8

Yesterday in Japan

A walk through the market at Nagasaki, the day after the *Albatross* arrived, probably changed the whole current of my life. There I saw great chunks of whale meat on sale for food. I didn't know that shore whaling was being carried on in Japan and I am sure no other naturalist did. Pacific whales were virtually unknown from a scientific point of view. If I could stay to study and collect specimens it would be a ten strike both for the Museum and for me.

The meat, I was told, came from stations in the Bonin Islands and the headquarters of the whaling company were at Shimonoseki. With Paymaster Van Mater of the *Albatross*, I went there the next day. Nothing could have been more cordial than my reception by the company officials. They were, they said, delighted to have an American scientist study at their stations and I could have all the skeletons the Museum wanted. The Bonin Island season was almost ended, but they had half a dozen stations in other parts of Japan. They would send me where I could find the most whales. So I said farewell to my shipmates on the *Albatross* while she started on her long homeward voyage across the Pacific. A few days later, I was settled in a tiny Japanese hotel in the fishing village of Oshima.

My life in Japan, thirty-three years ago, might have been lived in a different country and with people of a different race, judging by the present day Japanese. Seemingly the people I knew, and liked, have no relation whatever to the inhuman creatures we are fighting in this war of horror. It was only six years after their struggle with Russia when they were just beginning to emerge as a world power. The wine of success had already begun to make their rulers dizzy but dreams of Pacific empire and the "Master Race" for Asia had not yet permeated below the upper stratum of officialdom. The everyday Japanese was a likeable person, simple, full of *joie de vivre* and the worship of beauty. They were avid to learn of Western ways and the great world beyond their shores, but it was with the eagerness and humility of schoolboys. Their inferiority complex, the basic curse of the modern Japanese, then did not dominate their relations with foreigners as it does today. Their own ways and customs were good in their eyes. Since they were newly hatched from the egg of Oriental seclusion there was no reason to be ashamed of their lack of knowledge any more than is a child when it seeks information from its parents. Only when their politicians determined that their destiny lay in westernizing the race with frenzied rapidity, and they started to ape Occidental customs, did their inferiority complex begin to color their whole lives.

I watched them change year by year with amazing rapidity as they assimilated more and more Germanic ideas and *Kultur*. I saw them lose much of their courtesy and kindliness, their simplicity and charm. Each time I returned to Japan, there was less that was admirable and more of those characteristics which stamp the Japanese of today with the infamy of treachery and inhuman cruelty. But in those early days, when I went to live and study at the whaling stations, I had the feeling of being an honored guest of the whole community.

The language was a problem but I made up my mind that I'd have to learn Japanese. Armed with a grammar and a dictionary, I set out on a linguistic adventure. One gains a working knowledge of a new language surprisingly quickly if one wants food or drink or the necessities of life and must ask for them in the native tongue or go without.

The manager of the hotel was the only person in the village who spoke a word of English. With much laughter we laboriously read each other's sentences from our copy books if we had to use more than signs or simple words.

I had the adaptability of youth, but even then it took some doing to fit into the home life of a Japanese village. In the first place, there was no such thing as privacy. My little room fronted on the street. There were only sliding paper screens between me and the curious multitude. If I shut the screens, there would be a dozen wetted fingers making holes through the paper and a curious eye gleaming at every aperture. From dawn until the light was out at night, I was under intensive observation as though I were a specimen in an experimental laboratory.

Soon after reaching the hotel, the manager asked if I wanted a bath. I didn't particularly, but I knew I'd be branded forever as an "untouchable" if I refused. So I said yes, certainly, I'd love a bath.

"The-water," said my host, "is-clean. Only-five-people-have-bathed-already." He gave me a kimono and I proceeded to disrobe. I was lucky, for this particular bathroom happened to be enclosed. The water in the great wooden tub really was cleaner than might be expected after five baths.

I was just about to step in when the screen slid back and my host presented a smiling girl dressed in a lovely flowered kimono.

"This-girl-will-wash-your-back," he said, in his funny, slow English. With a yell I leaped for my bathrobe.

"Get out! I can wash my own back!" I shouted, pushing both of them through the door. Ten minutes later, he reappeared with a different lady.

"*This*-girl-will-wash-your-back," he repeated with an ingratiating smile.

"I'll be damned if she will," I roared and slammed the screen again. The next time he knocked timidly. His face was a picture of despair.

"If-you-do-not-like-my-servants, I-will-get-a-geisha," he almost wept.

Obviously I had committed a grave social error in not wishing to have my back washed by a young lady to whom I had not been previously introduced. "I'm in Japan," I thought. "Better do as the Japanese do. Bring on your girls."

The back-washing was performed to the accompaniment of many little bubbling exclamations in Japanese. Afterward, I learned their meaning. It was what in a modern girl would be called a "line." "How white your skin is," most of them ran. "Just like snow."

In less than a week I was laughing with my new Japanese friends at my strange conception of modesty as applied to such a natural function as that of bathing.

A public bathhouse fronted on the main street of the village, and I used to go there often for fun. Two great wooden tubs stood at the back above charcoal stoves. The rest of the room was a matted platform where the clients, both male and female, scrubbed and soaped themselves. Everything was wide open to the street. While the ablutions were being performed there never was conversation except strictly about the business in hand. It was quite all right to say to a girl whom you did not know, "If you'll wash my back, I'll scrub yours." She'd nod, give your back a thorough going over and then turn around while you went to work on hers. But this little personal transaction wasn't an introduction! Oh, no! You might meet her an hour later and she'd cut you dead. Just because you'd washed her back had nothing to do with the case.

I was intensely amused by the whole business and I don't mind saying that I never was so clean in my life as when I lived in Japan! It was very much the same when you came down to the common washroom to make your morning toilet. You didn't speak to anyone else until you had washed, shaved, and brushed your teeth. Then, when you were properly clean and ready to face the world, you turned to each person individually, bowed from the waist, and said, "Good morning to you."

The season was in full swing at Oshima and not a day went by without at least one whale. Because everything depended upon getting the meat to market in the shortest possible time, the whole operation of "cutting in" was unlike what I had seen in any other part of the world. The whale was seldom drawn out upon the "slip," but the blubber and meat were stripped off at the end of a long wharf while the carcass rolled over and over in the water. The moment it arrived men, women, and girls attacked it like vultures, cutting off huge chunks and loading them on fast transports. There were no regular hours and work never ceased until the last scrap of meat was on its way to market.

Often whales arrived in the middle of the night. I always went down to the wharf not only to do my scientific work but to watch the strange scene. Flares of oil-soaked waste lighted the station yard. Men and women, stripped to the waist, girls and children in blue kimonos or skin tight breeches, waded through pools of shining blood, slipped on the greasy blubber, and tore like demons at masses of steaming meat.

"*Ya-ra-cu-ra-sa,*" they sang in a meaningless chant as they strained and heaved at the colossal bones. The scene was weird and unearthly like a picture of the Eternal Pit with grinning devils at their business of torturing the Ungodly.

I worked as hard as anyone, and longer hours than the rest, for after the blood and grease of the cutting platform had been washed off, my notes and measurements must be transcribed while they were fresh in mind. Seldom did I have more than five or six hours sleep in any night.

I got on well with the men and that was most important. I must have been in the way a great deal but they accepted it good-naturedly. To get them laughing was the best method to gain time, and I purposely used to make mistakes in language. For instance, the Japanese words for "fish" and "cherry" are somewhat alike. I would call their attention to the beautiful "fish blossoms" or make other amusing substitutions. Shrieks of laughter greeted such remarks. By the time they had me repeat it once or twice, and had stopped roaring, my photographs and measurements were taken.

The problem of greatest scientific importance was to find out whether the whales of the Atlantic and Pacific were the same, or different, species and whether they migrated from one ocean to another. For future study, I had to take photographs, descriptions, and about forty measurements of each specimen and send skeletons to the Museum.

It sounds easy enough to say "I sent four whales to New York," but you can't just put a skull weighing twenty tons in your trunk and ship it off. Crates had to be made and I learned an important fact about handling Orientals. Tell them what you want and let them do it in their own way. It won't be your way, but that isn't important. After I got this idea through my head, the Japanese carpenter turned out crates which could have gone twice around the world. They were matched, grooved, and fitted like a cabinet job and he used very few nails.

I sent back four skeletons from the first trip: a sperm, sulphur-bottom, sei, and finback. The consignment weighed many tons. A humpback, two gray whales, a dozen rare and unknown porpoises, and two killer whale skeletons joined the parade to New York the following year. These gave the American Museum the finest collection of Cetaceans in the world. Most of them hang today in the Hall of Ocean Life.

From Oshima, I moved to the village of Aikawa, in the north of Japan, where the whaling company had one of their biggest stations. There I lived in a beautiful little doll's house overlooking the bay to a shore line fringed with twisted pine trees. They seemed incredibly old, and tired of life, as though they had seen too much history since the days of the Shoguns.

Also, I rated a servant girl of my own. Her name was Kinu. To look after me was her sole duty. Kinu was tiny and pretty and delicate and eighteen years old.

She looked like a beautiful butterfly when she fluttered about the house in a flowered kimono. I told her I liked it much better than the usual blue one and she said she did too and she'd like to wear it every day, but she couldn't because she had only one and it wouldn't last long. Then what would she do when she wanted to dress up? The next week I took her shopping at a town across the bay. There were three kimonos which she particularly liked and, as she couldn't make up her mind, I bought them all with *obis* (sashes) to match. Never did I see such transports of delight. From that time on she wore brilliant kimonos with reckless abandon, to the envy of all the village girls.

Kinu-san used to do the nicest little things besides cooking my meals, keeping the house immaculate, and waiting on me. Every morning, the screen would slide back softly and she would place a vase of flowers just where I would see them when first I opened my eyes. Also the *kakemona* (scroll picture) in the formal alcove of the room was changed every few days, for she borrowed new ones from her friends all over the village. If I were gone for several hours she would be kneeling at the door upon my return murmuring the polite phrases of greeting which, meaningless in themselves, nevertheless were pleasant to hear. "I have been inexpressibly lonely while you were away. The sun has been under a cloud. The hours were dark." That's the way they went.

After a few weeks at the station, I got some sort of an intermittent fever which knocked me out completely. There was no physician nearer than Tokyo, three hundred miles away, so I had to doctor myself. Every morning my temperature was subnormal, but by two o'clock in the afternoon it was up to 104 degrees. Sometimes I was half delirious. Kinu-san nursed me like a mother, keeping cold packs on my head, and hardly left my side to eat. She grew thin and tired.

I discovered, later, that every day she had gone to her little stone *joss* in a hillside temple and prayed that the American-san might soon recover. When I was well and strong again she asked me, shyly, if I would go with her to thank the *joss*. Not for the world would I have refused; so I walked up the winding path to the summit of the hill, clapped my hands three times to draw the God's attention, and thanked him solemnly that the fever had departed whence it came.

It is easy to understand why the Japanese have such a wonderful intelligence service if one has ever lived in a Japanese village. Curiosity dominates the race. They simply *must* know everything about you, to the minutest details, or it becomes mental torture. My house adjoined the station where all the officials lived, and I used to take delight in waiting until the end of the week when everyone was particularly busy making up reports for the head office; then I would start to unpack a trunk or suitcase. In a few minutes the manager would hop up to see what was going on; the assistant manager had to come, too, and all the secretaries. I would prolong every operation, taking out clothes, reading papers, and looking at photographs while each man practically had his nose in the trunk. All of them stayed till the bitter end and that was usually two or three hours. As a result, they would have to work far into the night to get their reports off on time.

They were imitative as monkeys. If I hung my clothes out to sun, every man jack of them hung out his clothes, too. I had a bottle of Eno's Fruit Salts on my table and used to take some in the morning. They asked me what it was, of course, and I told them "medicine." I noticed, nevertheless, that it disappeared like magic and one day saw a secretary, whom I couldn't abide, slide up and mix himself a big dose. Ah, said I, I'll fix you. Into the bottle I poured a powerful laxative, powdered, and awaited results. They came in due time. The secretary couldn't sit at his desk more than five minutes without leaping for the *benjo*. At last he said he was sick and asked me if I didn't have some medicine which would help him. Yes, said I, some of these fruit salts would be just the thing. He threw up both hands with a look of horror and rushed off to keep his next *benjo* appointment.

Within a few days, I knew every man, woman, and child in the village and by that time could speak pretty fair Japanese so long as ordinary affairs were under discussion. When I used to walk through the village at night, the greeting was invariably: "Good evening. Where are you going?"

Sometimes I'd give them a straight answer and they were satisfied. At others, I'd say, "I am going to Hell—would you like to come?"

"What! You are going to Hell? Where is Hell?"

Then I'd laugh and they'd laugh, too, in embarrassment, but you may be sure they'd tag along until I had reached my destination. Even on the streets of the big cities like Yokohama and Tokyo a foreigner was sure to be stopped frequently by students. They were studying English in school and wanted to practice. Always the conversations were: "How do you do? I am very well. What is your name? Where do you come from? Where do you live? How do you like Japan?"

In the earlier days, one could put them off with facetious answers or talk seriously, according to one's mood. But in later years, after suspicion of foreigners began eating at their vitals, I discovered that being humorous didn't work. Several times when some buck-toothed, bespectacled youth bored me with his questions and I gave him evasive answers, the police called at my hotel shortly afterward for a polite but thorough questioning.

There never was a doctor at any of the whaling stations, no matter how remote, and my popularity was enhanced because I had some medical knowledge. The Lord knows, I stretched what little I had acquired at the College of Physicians and Surgeons in New York far beyond its limits, but common sense and simple medicines did a lot. At least none of my patients died and I was confronted with some pretty tough assignments. The worst was amputating a man's hand, badly crushed in the winch, without an anesthetic. But it had to be done at once, and turned out to be a pretty good job. When the stork presented two mothers with babies, I officiated, assisted by the village midwife, who knew much more about how to handle the job than I did. Venereal diseases and eczema were the principal afflictions, but, fortunately, I had a plentiful supply of zinc ointment and potassium permanganate which worked wonders. I found, however, that all my patients insisted upon being given internal medicine, else they would not follow the external treatment. The more evil a concoction tasted the more certain they were of its effectiveness. I prepared a stock solution of colored water and quinine, bitter as gall, but harmless otherwise. It was amazing what cures it made.

Fourteen little ships were operating out of the *Aikawa* station and each carried a Norwegian gunner, for the Japanese had developed their shore whaling industry under the tutelage of the Norwegians, where shore whaling first began. They hired some of the best gunners to teach them how to kill and prepare whales, and purchased the ships and equipment in Norway. Later, as might be expected, they copied the ships and gear themselves and sent the gunners home. Some of the Norwegians I had met in other parts of the world. They were rough men, but we had a lot of fun together. Usually at night half a dozen of us gathered at a village tea house for dinner or a geisha dance. I used often to go out with one of them for a week or more at sea to study and photograph live whales. In August, I had an experience which came near to being the end of the road for me. When we got ashore, I wrote the story in my journal.

The harpoon struck a big finback between the shoulders and as the bomb didn't explode he was virtually uninjured. Dashing off like a hooked trout he took rope so fast that the brakes on the winch were smoking. Cable after cable was spliced together and before his rush could be checked he had out nearly a mile of line. Then the brakes were set and he towed the ship forward with the engines going at full speed astern. After an hour of this, even his great strength began to fail. The rope was slowly reeled in, but we could never get closer than about half a mile. Then a wild rush would take him off again in a smother of foam. The fight dragged on for six hours.

"He'll keep us fast all night," said the captain. "I'm fed up with this. I'll send a boat and lance him."

"Let me go," I asked. "I want to get some close-up photographs. I'll pull one of the oars."

"All right. You'll take the *praam.*"

A *praam* is a Norwegian boat big enough for three or four men, which sits deep at the stern and can be spun around almost like a top. The mate carried a long slender lance. A seaman and myself were at the oars. The whale lay at the surface, nearly a mile away, now and then blowing lazily.

As the tiny boat slipped up from behind, the body loomed bigger and bigger. To my excited eyes it seemed like a half-submarine. Actually, the whale was seventy-two feet long. Standing in the stern, with lance poised, the mates steered us up right beside the whale.

"Way enough," he whispered.

Swinging the *praam* about, we backed up till the boat actually touched the gray body. Bracing himself, he plunged the slender steel deep into the animal's lungs. As his arm went down, we gave a lusty pull at the oars. My right oar snapped short off and the *praam* swung directly against the whale. Up heaved the great body. I saw the tail, twenty feet across, weighing more than a ton, waving just above my head. It appeared to hang in mid-air and then to be coming down right on me. Never will I forget that second—it seemed hours long! The tip of the fluke missed me by six inches, but the side of the boat was smashed like paper.

I was in the water with the other two men, swimming for the floating wreckage. We caught hold of the stoved boat and looked around. The whale lay at the surface a few fathoms away, blood welling from the blow holes. We could hear the rattling of the winch as it wound in the line, while the ship crept closer. Suddenly I felt something bump my foot. It was a huge shark. The water was alive with flashing white bellies and great sharp fins. I was yelling like mad and absolutely sick with fright. The other two men joined the chorus. They were as scared as I. Wrenching off pieces of wood from the smashed boat, each of us hung on with one hand, beating the water with the other. But the sharks weren't really interested in us. They swarmed like flies about the dying whale, drawn by the blood pouring out of its nostrils.

The ship came alongside and the captain shouted:

"Hang on a little longer. I want to kill this whale before I lower a boat."

With that he calmly left us and went over to the dying finback. We yelled and cursed while the whale was drawn up under the bow and lanced again. The captain only grinned. When we were on deck I walked up to him, mad as a hornet.

"What the devil did you mean by leaving us there among the sharks while you killed that damned whale?"

"Well, you were all right, weren't you? The water is warm and you had a nice boat to hang onto. As for those sharks—shucks, they aren't man eaters!"

I wanted to shoot a whale myself, but the gunners weren't very keen about it and I wasn't surprised. It wasn't sport to them. With a thousand dollars blowing right in front, you don't care to have the package sink back into the briny deep through the bungling of an amateur. But I was determined to add the biggest of all animals to my game list; moreover, I'd seen a lot of whales killed and was sure I could do the job with neatness and dispatch. So I agreed to pay the value of the whale if I got a fair shot and missed.

Theoretically, I did know all about it, but a psychological factor was involved for which I wasn't prepared. A whale doesn't rise vertically. He comes up obliquely, the great flat head emerges, he shoots out a column of vapor from the blow holes and then revolves slowly in the arc of a circle. Just when the back is at the top of the curve is the time to fire, for then the harpoon crashes into the heart or lungs. It was a big finback that we were after. The sea was as smooth as a mill pond and I stood at the gun, every muscle tense. Suddenly a tell-tale "slick" showed on the water and the gunner yelled, "Get ready. He's coming." I could see the huge body rising like a gray ghost and had the sights lined up as he burst to the surface. Fifteen feet of head showed right under the bow. I knew it wasn't time to shoot, but somehow pressed the trigger. The harpoon struck him on the back of the head and glanced off. A whale hasn't a very big brain, but a hundred-pound bar of iron, propelled by three hundred drams of powder, hits quite a punch and it knocked him out for a count of about two hundred. He rolled over and lay still while the men worked frantically to reload the gun. Just as the huge beast was beginning to notice things again, I let him have the second harpoon right under the flipper and he went down for good. It came near to being a very expensive morning for me.

The finback was hauled to the surface, blown up by an air pump from the engines, and left to drift, marked by a flag. We went off to hunt a pair of sei whales which the man at the masthead reported only a few miles away. I had lost so much "face" in my first try that I begged for another shot under the same conditions. This time, I had myself under better control and did a very creditable job sending the harpoon smack into the heart. The bomb on the tip burst and the whale died without a quiver.

When the station at Aikawa closed in August, I had eighty tons of specimens crated and at last got them aboard a freighter bound for Yokohama. Then I tagged along to make final shipping arrangements on a vessel sailing directly to New York. Besides the great haul of specimens, my notes and photographs held more information about whales, living and dead, than had been gathered since whale hunting first began back in the dawn of history. All this, strange as it may seem, because I was the first naturalist who had ever had the opportunity of studying Cetaceans in the Pacific with the unrivaled opportunities which the Japanese shore stations afforded. It was an absolutely virgin field and I happened along with the youth, enthusiasm, time, and money to take advantage of the opportunity with which my Lucky Star presented me.

Chapter 9

Strictly Personal Explorations

I had been a year in the Far East when the whaling season ended but still I didn't want to go home. Casting around for some excuse to stay in Japan, I thought of fish. Shimonoseki was the headquarters for a fleet of trawlers that worked on both sides of the island, in the Inland and Japan Seas. The Museum's department of fishes would like a collection from Japanese waters, I was sure, for it would cost them very little. So I sent a persuasive cable, asking for only three hundred dollars and guaranteeing fish by the score. It worked. Within a week, I was on my way to Shimonoseki.

The ports of China and Japan in those days were pretty wild. Business was done casually when it did not interfere with pleasure. Shimonoseki rejoiced in the reputation of being the hardest drinking port in the Far East. Only about twenty foreigners lived there and they were all older men. Most of them were English, presided over by the genial British Consul. A German agent and an American doctor represented the only other nationalities.

It wasn't easy to be just a moderate drinker in the Orient and especially difficult in Shimonoseki. As a rule, most of the foreigners gathered at the club about one o'clock. As each man came in the door he said to the boy, *"Mina kita,"* which meant "Ask everyone what he will have to drink." If there were ten members in the club, ten drinks were offered and a lot of men took them all. As a result, the servants "poured" the gentlemen into their rickshaws in the middle of the afternoon and their offices saw them no more that day.

I was perfectly willing to have my share of fun, but I wanted to take it in my own way. I had a job to do and my self-respect to keep and I didn't intend to offer either as a sacrifice at the bar of the Shimonoseki Club. Therefore, instead of living at the hotel, I got myself a house a mile out of the city where I could be alone. It was a sweet little place, set high up in a pocket of the hills overlooking the entrance to the Inland Sea. I rented it for thirty yen (fifteen dollars) a month. This also included a man, his wife and daughter, a cat, four gold fish, and a frog. The Japanese lived in the back room; the cat and I had the front of the house; the gold fish and frog resided in a pool of the garden spanned by a tiny red lacquer bridge.

I was extraordinarily happy there. Every afternoon I watched the sun set over the Inland Sea behind the gnarled old pine trees, and the shipping in the Straits. Junks with bat-winged sails, square-rigged fishing boats, and stubby trawlers. Sometimes a huge ocean liner slipped through the narrow channel and I could hear music and see men and women dancing on deck. As darkness gathered, the strumming of a *samisen,* feminine laughter, and the soft plaintive notes of a bamboo flute floated up from a tea house below the hill. Then my frog began his evening serenade.

Every morning I went first to the docks where the trawlers brought in their fish. I looked them over and then proceeded to the general markets. The old women at the stands soon came to know me well and saved every unusual specimen. Some even gambled by buying strange fish that were not very good to eat in hopes that they might suit my fancy. I always took them, of course, even though they were not new to the collection. The sharks were particularly interesting and of great variety. The fins and tails were dried for export to China, where shark-fin soup is a national delicacy. The ordinary fish only had to be injected and dropped into a tub of formalin but the sharks must be flensed. I wore the tips off all my fingers on the sharp tubercles of the skins.

Both in China and Japan, carpenters used dried shark skins for sandpaper.

Finally, I had to admit that there was no excuse for me to remain longer in Japan. I had collected every species of fish that the trawlers and market produced, and even my fertile brain could not conjure up another legitimate project. So I cabled the Museum that I was on the way home but not to expect me too soon. My salary of a hundred and fifty dollars a month had been piling up in the bank and I was minded to spend it seeing the great world. It could best be done by being unencumbered with baggage for I had decided to return by way of Suez and the Mediterranean.

As long as I was in the tropics, whites were all that was necessary, so I took only one cloth suit and a tuxedo. Two suitcases were ample. When I got out of hot weather the whites were mailed to New York and their place taken by another suit purchased en route. Since I was going to wander where the whim of the moment carried me, I could hop off a ship or a train at a moment's notice, for my baggage and myself were virtually one. I've always hated too much planning in advance on a pleasure trip for one never can tell what one is going to find. The only definite project in my mind was to see all the biggest natural history museums of Europe. That, of course, meant Italy, France, Germany, and Belgium, ending up in England where I wanted to do some studying at Cambridge University. Otherwise, I was footloose and fancy free.

Curiosity about life has always been my dominant characteristic. I never can learn by someone else's experience; I must try it for myself. I had read about all sorts of temptations that the world had to offer. Some of them sounded exciting; some didn't. Anyway I was keen to try everything once and see what my personal reactions would be; to find out what sort of chap I really was inside. I don't think that I phrased it mentally but I did know that I was undertaking an exploration of myself. It was just as exciting as when later I stood on the top of a mountain in Central Asia looking out over a land which no white man had ever seen before and wondered what hidden secrets it contained. Thus my strictly personal adventure began.

Fiction writers had led me to expect romantic and glamorous personalities, refined vice, and exotic surroundings in the Oriental underworld. I visualized beautiful but incredibly wicked sirens reclining on satin-covered divans amid priceless embroideries, jades, and porcelains, wreathed in an aura of intoxicating perfumes. Sirens, who lured men to destruction with sensual pleasures undreamed of in Western minds. I visualized clever women who ruled their world like czarinas, and pulled the strings which made puppets dance even on the stage of international politics. But I found nothing romantic, nothing glamorous. All of it was a figment of the imagination. Opium I tried one night. It made me sick. Ah, said they, you must get used to it. Try it again. What you will experience is beyond the dreams of man! So I tried it again and got sicker still. To this day the sweet smell of opium smoke makes me violently ill.

But it wasn't time and money wasted. I was learning about myself. I discovered that I was too fastidious, had too much self-respect, ever to fear the best that the Oriental underworld had to offer. A very intriguing dream was shattered, but since I had learned that it *was* a dream and existed only in the imagination of fiction writers, I could dismiss it from my mind. It was a job well done.

Like everyone who lives in the Orient, I had seen the terrible effect of drugs on men and women of every nationality, and I wanted to find out *why* it made them slaves. I discovered that drugs were something I couldn't fool with. Cocaine, morphia, heroin, hashish—I tried them all to satisfy my insatiable curiosity. And what a jolt I got in my personal exploration! Alcohol, I had long since learned, was something I could take or let alone at will. I enjoyed a Scotch or a cocktail, but only for congeniality's sake. Very seldom did I ever take a drink by myself. For months at a time I never touched a drop of liquor.

Opium, as I have said, was like an emetic to me, so I tried cocaine. It took effect within the hour. I got, I suppose, what alcohol gives to some men: a sense of the most wonderful well-being, a feeling of power unlimited, and the ability to do anything in the world. Morphia does the same.

And so it is with any drug. I find it difficult to forget the heavenly exhilaration. Two or three times I have had cocaine for minor operations and the effect is always devastating. The doctors have said that no patient they ever had reacted so tremendously to narcotics. Knowing that this is my weak spot, I avoid it like the plague. Never will I let a physician give me drugs of any kind if it can possibly be avoided. I could become a slave to cocaine or morphia within a week.

From Japan and China I wandered slowly southward and eventually reached Singapore, the "Cross Roads of the Orient." It was a strange human cocktail, basically Malay but with a dash of China, India, Burma, England, and Holland to enliven the mixture. I would have loved it except for the damnable heat. Borneo and the Dutch East Indies had demonstrated that I am not a hot weather man. It was all right, of course, as long as I could sit on the wide veranda of the Raffles Hotel sipping iced drinks, but my physical motor seemed to run out of gas in the wet, torrid heat. I always felt full of push and go in the north but never in the tropics.

I stayed for a while in Singapore, exploring the town both by night and by day without an incident of sufficient importance to linger in my memory. I was getting rather bored, waiting for a ship, until I discovered a whale skeleton and some rare porpoises in the Raffles Museum. My excitement mounted as I searched their study collections and I had a wonderful week reveling in pure science. I sailed, with a notebook full of new facts and data and a sense of great satisfaction, for Kuala Lumpur and Penang, there to gorge myself on mangoes and mangostenes and the world's most delicious curry.

Colombo, on the beautiful island of Ceylon, I remember chiefly for the ebony elephants and moonstones which one bought by the cup and gambled on finding one among them worth more than the price one paid; for Mount Lavinia, outside the town, where I ate delicious sea food and swam in the white foam of the surf washing golden sands. But particularly I remember a garden, straight out of fairyland, behind the Grand Oriental Hotel. The picture is as vivid in my mind today as when I entered it thirty-three years ago. The tip of every palm frond bore a colored light, and an orchestra of strange native instruments played under a natural arch of tropic flowers. There was a fountain splashing scented water into a rock basin of rose and green and tables heaped high with flowers, each under its own little arbor of fragrant blossoms. This, I thought, is where I should find the romance of my dreams, but alas, there was no lady! Only two British Army officers, rather dull, who talked of Oriental politics when I wished not to speak at all.

The Suez Canal was intensely interesting. It was incongruous to be on a great ship, moving through a narrow lane of water no bigger than a creek, with the desert on either side. There was another vessel ahead of us and where the canal made a sharp turn she seemed to be sailing through the sands without benefit of water. Even though the heat was blistering, I stayed on deck all day long.

At Port Said I ran into trouble, or what might have been trouble, had it not been for a friendly Arab. It was in the middle of the day when we landed and, from the moment I stepped ashore, there was a horde of native procurers dogging my heels whispering what they thought were seductive phrases of the beauty of their clients: Arab, French, Egyptian, Armenian, and girls of every nationality under the sun. They were like a swarm of flies; you brushed them off but still they buzzed about your ears. Finally I got rid of all but one. He was a cringing little beast, Arab, I think, and hung persistently at my elbow until I lost all patience. Finally I turned on him and snarled, "Get out, you, and let me alone or I'll knock you down." Instead of leaving, he put his filthy hands on my arm and stuck his face almost into mine, frantically gurgling his obscene patter. I whirled, brought my right fist up in a short jab squarely on his chin, and he went down like a felled ox. His head struck the curbing and a trickle of blood spilled out over the stones.

Instantly, it seemed, the street was alive with screaming natives. Where they came from was a mystery; they seemed to spring from every building and through the very earth itself. I backed into the doorway of a closed shop and faced the mob. They stood glowering, waiting for someone to rush me first. Suddenly, I heard bolts being drawn, the door to the shop opened, and I was pulled backward into darkness. The door slammed shut, the bolts clicked again, and a white-gowned native said in perfect English, "Hurry. Come with me. They will kill you if you don't."

In the half darkness we ran to the back of the room, out another door, down an alley, and up a short flight of steps into what seemed to be a warehouse. Stumbling over boxes and bales, my guide took me across the building and down a narrow street at the end of which I could see the gleam of water.

"Your ship is there at the left. Walk, *don't run,* and get on board."

I tried to thank him but he pushed me forward, smiling.

"That's all right. I saw it from the shop window. I'm a student and I like Americans. Don't go off the ship again. They'll hurt you if they find you alone. Those are bad people, bad people, the worst in Port Said."

"Do you think the man I hit is dead?" I asked.

"I don't know; maybe so. If he is, the police will be glad. But I think he would be too hard to kill."

I got back to the ship in short order. Fortunately the captain was on board, and I told him the story of my adventure.

"They're a bad lot," he said. "We've had one or two other experiences somewhat like yours. You were lucky to have got out of it alive."

"What," I said, "will happen if I killed the fellow?"

"I don't know. We'll just have to wait and see."

The incident ended there so far as I know. I had already decided to leave the ship and have a look at Egypt. That night the captain and two other officers saw me safely on the train for Cairo.

Egypt delighted me. I engaged my own little caravan and traveled into the Fayum just to get the feel of the desert. White nights in the moonlight, sleeping on the sand near grunting camels, the stately natives, and pavilion tents seemed to have transported me into the stories of the *Arabian Nights*. That short trip showed me that the desert was what I could love most of all.

After Egypt there was Italy, where at a moment's notice I hopped off the train in the Tyrol with my two little suitcases and walked across the Brenner Pass to Austria. Then Germany, Belgium, and France. Everywhere I had adventures, small ones, to be sure, but exciting to me for I absorbed each new experience like raindrops falling on dry earth.

England meant only Cambridge University and the British Museum of Natural History. Mr. Oldfield Thomas, one of the world's most distinguished mammalogists and Keeper of Zoology, offered me a position as his assistant. It pleased my ego but I told him I was wedded for life to the American Museum of Natural History. How completely that was true I hardly suspected then.

When my money was entirely gone I sailed on the *Kaiser Wilhelm der Grosse* for New York. After paying the tips on the ship there was exactly five cents left to take me and my two suitcases to the American Museum.

Chapter 10

Korean Devilfish and Killers

When I was in Japan I heard of a strange whale called the "Koku Kujira," or "devilfish," which was being taken off the Korean coast. The description sounded much like the California gray whale. But the gray whale had been extinct for fifty years, at least as far as naturalists were aware. The animals used to come into the California lagoons to breed; there they were hunted ruthlessly, and eventually disappeared.

I thought a good deal about this Korean devilfish during that year in Japan. It was either a species new to science, or else the lost gray whale. If I could go back perhaps I would have the opportunity to rediscover a supposedly extinct species! When I reached New York plans were ready in my mind to return to the Orient. At my first luncheon with Professor Osborn, I told him how necessary it was to capitalize upon our cordial relations with the whaling company and get the devilfish at once. It wasn't difficult to convince him. It never was, when science could be advanced.

The devilfish appear twice a year along the Korean coast, as they go northward in the spring, and again in the autumn upon their return journey to warmer waters. Therefore, I planned to reach Korea in January to be ready for the spring migration. It meant that I would have only a few months in New York, but that would be quite enough. I knew that I would be ready to seek the open road again before the year was out.

The next months were busy ones. I resumed my interrupted studies at Columbia University, described several new porpoises from Japan, and was appointed an assistant curator of the Museum.

The only trouble was that I never had time to "curate." In order to curate properly a curator must stay at home—at least some of the time. I couldn't do it. During all my years at the Museum I almost never returned from an expedition without having plans ready for another. I had found it wise to strike while the iron was hot. The enthusiasm of returning from a successful trip carried great weight in the plans for a new one. Then again, I was afraid I would get so immersed in Museum affairs that the authorities might think I had better stay at home for a while. If the new trip had been approved, and was in the offing, it was much easier to keep my particular decks clear for action. Not that I didn't like Museum work. On the contrary, I loved it. The Museum had come to be a part of myself, and there was no phase of the activities in the great institution that did not fascinate me. But I loved wandering more. Sometimes when I walked across the park on a starlit summer night I used to look up at the drifting clouds, going with them in imagination far out to sea into strange new worlds. Then I would count the days that still remained before I could set my feet upon the unknown trails that led westward to the Orient. Seldom did I leave my office in the Museum before one o'clock in the morning. I had so much to do and so few hours in which to do it that I hated to waste any of them in sleep.

An especial stimulant to my desire to go to Korea was a book by Sir Francis Younghusband on his first expedition in 1879 to the "Long White Mountain" near the Manchurian frontier. He wrote of the wonderful fields of wild flowers on the slopes of the mysterious mountain; how its summit was white, not from snow, as supposed, but from pumice; and of the beautiful lake in its crater. But what intrigued me most was when he related how he and James and Fulford had looked across the vast forests of Korea, through which no white man had ever passed, and wondered what secrets they contained.

I wrote to Younghusband. He replied that the forests were still unexplored, and that if I could go in from the Korean side and make a traverse to the base of the mountain it would be a really worthwhile job of exploration. It fitted in beautifully, for I would catch my whale during the early spring and have the whole summer available for the Korean expedition. Professor Osborn agreed enthusiastically—provided I could raise the money. The Museum had funds only for the whale part of it; the rest was up to me.

By that time, I knew a good many people of wealth in New York and I set about selling the expedition. The price was ten thousand dollars. I promised a collection of mammals and birds which would be extremely valuable since the Museum had not a single specimen from Korea. Eventually, I got the money. Now I was ready to begin, at last, the program of land exploration which had been my ambition for as long as I could remember.

I picked the maiden voyage of the beautiful *Shinyo Maru* for my return to Japan. We landed at Yokohama, on December 31st in the afternoon. I went up to Number Nine to bring Mother Jesus a present from New York. She was glad to see me. That night she invited me with two of my friends for a New Year's Eve party. We had a room opening on the main court which was a mass of flowers, lanterns, and bright hangings. There, geishas danced while blind flute players strolled about from room to room. Mother Jesus was as busy as a bird dog, but she found time to tell me the news of what had happened in the Orient during the last six months. As a return gift, I was presented with a beautiful silk kimono which for many years I used as a lounging robe.

She wasn't very happy about the political situation. Every day, she said, the government was making new laws tending to restrict personal liberty. They had sent her a whole list of regulations. After a while, if things went on, Number Nine would lose all its character and gaiety. There was, however, plenty of fun and laughter that New Year's Eve and I was as excited as a child at being back in the Far East.

A few days later, I sailed across the Japan Sea to the whaling station at Urusan, set in a picturesque bay among treeless scrub-covered hills. All the employees were Japanese, but Koreans swarmed along the dock as the ship pulled in.

That very night, the station whistles roared out the call of whales. Flares, appearing like magic, threw a fitful light over the long dock and the black water of the bay. In ten minutes, I had drawn on my long boots and heavy coat. It was bitterly cold outside. The whaleship *Main,* shrouded in ice from stem to stern, swept proudly into the bay and slid up to the crowded wharf. From her bow drooped the huge, black flukes of a whale. My great moment had come. Either I was to find a species new to science, or rediscover one that had been lost for half a century!

First glance showed that the flukes differed in shape from any that I had ever seen; also they were marked with strange gray circles. When the cutters hacked through the body and the posterior section came slowly into view, I saw that the back was finless and the dorsal edge strongly crenelated. Up came a wide, stubby flipper; then a short arched head. These told the story. It was the gray whale, beyond a doubt.

When the last bit of meat was on the transport, I went to bed, but not to sleep. I was too excited. After vainly trying for half an hour, I lit a candle and spent the rest of the night checking my observations. Two more whales came in the next day and there was ample time to take photographs and measurements. That night I was so tired that even an invitation to a geisha dance and dinner at a tea house in the village could not tempt me from the *futons* on the floor of my paper-screened bedroom.

During the next six weeks I examined more than forty gray whales; saw them hunted by men and killers; learned their clever tricks to avoid their enemies; and pieced together, bit by bit, the story of their wanderings. The days at sea were torture. Always heavy weather and deathly seasickness for me, bitter cold, sleet and ice. Standing behind the gun for hours on end, my oilskins stiff from frozen spray, I used to curse the sea. Why had I deliberately chosen a job which took me off the land? But hardly was I back on shore transcribing my wealth of new data before the suffering was forgotten and I was keen to go out again.

One day a herd of killer whales put on a fascinating but horrible show for us. Killers are the wolves of the sea and hunt in packs. Armed with a double row of tremendous teeth they will literally devour a whale alive. We were chasing a big gray whale about fifty feet long close in shore where he was trying to escape by sliding behind rocks. Suddenly, the high dorsal fins of a pack of killers appeared, cutting the water like great black knives as the beasts dashed in. Utterly disregarding our ship, the killers made straight for the gray whale. The beast, twice the size of the killers, seemed paralyzed with fright. Instead of trying to get away, it turned belly up, flippers outspread, awaiting its fate. A killer came up at full speed, forced its head into the whale's mouth and ripped out great hunks of the soft spongy tongue. Other killers were tearing at the throat and belly while the poor creature rolled in agony. I was glad when a harpoon ended its torture.

By the time the station closed in March I had boxed and shipped two gray skeletons; one for the American Museum and the other for the Smithsonian Institution in Washington. Then I went up to Seoul for the expedition into the northern forests. Seoul, the ancient walled capital (the Japanese, with their mania for changing names, call it "Keijo") was like a western mining town in an Oriental setting. A concession for the Chicsan gold deposits had been granted to American interests. There were also iron, coal, and tungsten. Saturday nights, the miners came to town to celebrate, gathering at Sontag's Hotel and the Club. I met a dozen American engineers. Among them Eddie Mills of Harvard, John Francis Manning, and Larry Farnham. Sontag's was a hotel where you did as you pleased and no one said you nay. My friends sometimes staged wild parties. But it seemed to be expected and the miners paid for the damage, of which there was often quite a lot in the way of broken tables, windows, and dishes, without question. The life was free and gay and careless, just like any mining town in America where money came easily and there weren't many ways to spend it.

I saw a lot of the Koreans in the months I spent there and came to like, as well as to respect, them. True, their corrupt government and the tyranny of the nobles, or *yangbans,* had made them into a nation of inveterate loafers, but that was largely the fault of their social system. A *yangban* could not work else he ceased to be a *yangban,* and his power over the common people was almost absolute. If a peasant had acquired a sum of money, a *yangban* was certain to appear at his door and request a loan, politely to be sure, but it was a command nonetheless and the peasant dared not refuse. Therefore, the common people beat the game by working just enough for the needs of each day, letting the morrow take care of itself. It seems to me that the Korean *yangbans* strongly resembled our income tax collectors.

But the Korean is inherently a gentleman and respects himself as such. He may sit in the sun all day, smoking his long pipe and talking about the evils of his particular world, albeit doing nothing except talk, but there is about him an unhurried serenity which I found delightful. Probably it is because I am neither unhurried nor serene myself that I liked it so much.

Dignity, they say, is a matter of inner consciousness and the Korean must be full of it; otherwise he couldn't look as dignified as he does in his comic-opera costume. Atop his head is perched a sugar-loaf hat of woven horsehair, in shape the reduced replica of our Pilgrim Father's head gear. Ribbons, tied under his chin in a fetching little bowknot, keep it on, but it is always tilted at a Happy Hooligan angle over his ear. A flowing gown of white cotton, resembling nothing so much as an old-fashioned nightshirt, reaches to his knees and covers a pair of baggy, heavily wadded trousers which bunch up over his rear like a tied gunny sack. Purple ribbons wrap them about the ankles. Only a married man can wear a hat and do up his hair in a knot which shows through the horsehair fly-cage like a walnut. Only a married man, too, is considered to be a "person," or has any standing in society. Even though a bachelor may be sixty years old, he is still referred to as a "boy" and has to wear his hair in a braid hanging down his back.

The Korean woman, according to our conception, appears to have her ideas of modesty slightly mixed. From the waist down she looks mid-Victorian in a voluminous skirt. Topside is where quite a bit is missing. A diminutive jacket buttoned tightly about her neck allows not even a peek at her throat but leaves her breasts entirely bare. From a score of years of wandering about the world amid the beauties of many lands I rather fancy myself as a connoisseur of pulchritude. It is my considered opinion that the Korean girl, by and large, has more to offer than any other lady of the Orient. Her little oval face, cream-white skin, soft brown eyes, and cute little nose would make the judges in any beauty contest consider carefully before marking up their charts. When she walks along the street custom decrees that her eyes must be modestly cast down and she should look neither to the right nor to the left. Be that as it may, there is very little that she doesn't get. She is intelligent, too, as I learned from experience, though that is not easy to come by. Theoretically a "nice" Korean girl is kept in the seclusion of a nun in a convent. But there are ways and ways and girls will be girls. Even their own people are supposed to see little of the neighbor's women folk because every house is surrounded by a wall and the female members seldom go beyond the domestic confines. But even walls can be scaled by both sexes and the hot blood of Korean youth finds ways to "sit under the apple tree."

I liked both Korea and its people. There was something mysterious and intriguing about the country in which, until the Russo-Japanese War of 1904, the king had maintained a policy of rigid exclusion. Korea was called the "Hermit Kingdom" and no one knew much about it except that wonderful brass-bound chests and lovely bowls sometimes found their way out; and there were tigers that lived in caves, and ginseng, the root shaped like a little man, for which the Chinese paid exorbitant prices because it was supposed to be an aphrodisiac.

But after the war the country was mercilessly exploited by the Japanese and during the first years the worst riffraff from across the Japan Sea sucked the last drop of lifeblood from the expiring kingdom. Then, "for the good of the country," as I was told by numberless Japanese officials with a shake of the head and a sanctimonious smirk, Korea was formally annexed in 1910. For the good of the Japanese it was, because even though they did build roads and railways and develop the country, they engendered a lively hatred in the conquered people which was comprehensively expressed by a Korean cook I took on my expedition into the interior.

"I tell you, Mr. Andrews," he said, "no matter what people say, way down in his stomach every Korean hate the Japanese and always will."

That the little yellow men know this, and fear it actively, is shown by the slaughter of Koreans during the earthquake of 1923. Some hysterical ape-man started the rumor that the Koreans were revolting. The word spread like wild fire and every Korean man, woman, and child within their reach was butchered like sheep in the stockyards.

Chapter 11

The Long White Mountain

In Seoul there wasn't much that I could learn about the northern forests, because no one had been there. But I did get a Korean cook and a Japanese interpreter. He was assigned to me by the Foreign Office because the American Consul said it was better so. When he turned up at the hotel for our first interview, I was astonished. He wore an ancient frock coat and a badly ruffled top hat much too big. I explained that we were going on a trip of real exploration where we wore rough shirts and high boots, not long coats and silk hats. That did not deter him. As a matter of fact, he proved to be an excellent man. He stuck to me when things looked pretty black, and I developed a real affection for him.

There was one experience which I am not likely to forget. It happened after we had gone in a hundred and fifty miles from the village of Seshin on the west coast just south of the Manchurian border. Near the last outpost of Japanese soldiers a man-eating tiger was reported to be ravaging the countryside. Already he had killed half a dozen children and the villagers were in despair. They welcomed me with open arms. Would I kill the tiger? There was nothing I was more eager to do, and they recommended a fine old Korean hunter named Paik. He had received the honorable title of *Sontair* because he belonged to the fraternity of hunters who had killed a tiger singlehanded. It was something like an English knighthood, for public service.

The animal's custom was to go from village to village picking up pigs, chickens, dogs, and now and then a fat Korean child. Never did he stay more than a day or two in one place. A breathless native would arrive at my camp saying that the tiger had been seen at his village twenty miles away. In fifteen minutes, Paik and I would be gone. Always we drew a blank. In a day or two, news would come that the tiger had appeared somewhere else. We chased him up hill and down, till my shoes were thin and my patience thinner.

Then one day a native reported that the big cat had just gone into a cave near the top of a mountain only a few miles away. Paik and I were there in an hour. Sure enough, fresh pug marks showed in the dust at the entrance. We sat down to wait, concealed behind a clump of bushes. The sun set and it grew dark. Still, there was moonlight and the tiger couldn't get out of the cave unseen. It was a long night. In the first gray light of dawn, we examined the dirt in the cave's mouth. No new tracks.

"If he won't come out, we'll go in and get him," announced Paik in a matter-of-fact voice. Just like that! "We'll go in and get him!"

I hadn't lost any tigers in that cave! Besides, the poor old tummy needed food; I was tired and, most of all, scared pink at the thought of crawling into that cave with the tiger sitting there comfortably ready to receive us. When Paik said, "Are you going?" I gave him an evasive answer. When he urged me, I was more explicit. I told him to go to hell. He stared. "You've got a flashlight. The tiger won't charge fire. He'll be frightened." Well, I was frightened, too—awfully frightened.

"Besides, I'll be behind you with my spear," announced Paik as though that made everything all right. I gave him a still more evasive answer. Then he lost all patience.

"If you are afraid, give me your rifle and I'll go in alone," he said, but the look he gave me told just what he thought about white men who pretended they wanted to kill a tiger. That was a bit too much.

"Let's go," I said in a husky voice.

We had to crawl in on hands and knees for the cave was less than shoulder high. I went first, rifle in one hand, electric torch in the other. Paik followed with spear advanced. As a matter of fact, it was advanced so much that it kept pricking me in the rear end to my intense annoyance. I wouldn't swear that he did it on purpose, but I suspect him strongly.

About twenty feet from the entrance was a small chamber. A sickening smell of rotting meat almost suffocated me, but no tiger. We could see that the passage veered sharply to the left. I felt morally certain that the damned cat was waiting just there ready to reach out and claw me when I turned the corner. Paik seemed to know I had lost my nerve and gave me a few extra prods with the spear. Sticking my flashlight out at arm's length, I edged around the rock. Thank God, no tiger! The passage led on dipping slightly downward. In the distance, I saw the faint gray of daylight and knew that the tiger was not at home.

We emerged on the other side of the peak in a deep gorge filled with boulders which concealed the mouth of the cave. Pug marks plainly showed in the soft sand. They were all leading outward. Evidently the animal had scented us while we were watching the entrance to the cave and had quietly slipped out the back door. The natives did not know there was an exit to the passage for they had such fear of tigers that no one would go near that part of the mountain. We heard of the beast next day at a village twenty miles to the south. After three weeks, I gave up for I had to get on with my main job.

The traverse through the larch forests to the base of the Long White Mountain was a mixture of elation, discouragement, utter exhaustion, and final satisfaction. I had four Koreans and eight ponies. The men didn't want to go for the Koreans are timid; they are not the explorer type. It was only because the Japanese gendarmes ordered them to do so that they reluctantly consented to accompany me. Never having been more than a few miles into the forest, to the point where the swamps began, they were certain we would be lost and die of starvation. My compass, of course, they could not understand. It was a pretty grim place, I must admit. I killed a bear on the fourth march, but after that, for many days, we did not see another sign of life. The forest became denser at every mile with more swamps and surface water. Time after time, our ponies were mired and had to be pried out of the mud with poles. Lush ferns and rank grass made walking difficult. The trees were festooned with long streamers of gray moss which formed a thick canopy overhead. Down where we were, there was only a gloomy half light occasionally shot through with patches of thin sun. No sounds broke the stillness except the calls of the men. No birds or animals; not even a squirrel. To make matters worse, it began to rain. Not a hard refreshing rain, but a dull drizzle which continued for a week.

The men were completely disheartened, frightened at the gloomy stillness of the forest, and exhausted by strenuous work. They began to talk furtively among themselves and when we camped were ominously silent if I passed near their fire. The interpreter said they were planning to desert that night with the ponies and food, leaving us to die or get back as best we could. We were only two days' march from the base of the *Paik-to-san* and I determined to complete my traverse against all odds. I told the men that we must reach the mountain; that I would give them double wages; further that I should watch all night and if anyone touched a horse he would be shot without mercy.

They didn't like it much. My ultimatum was received in silence. The interpreter and I stood guard by turns through the night. Now and then one of the men got up to replenish the fire but they made no move to leave camp. The next night was a repetition of the first. Both the interpreter and I were exhausted from lack of sleep and hard work. We wondered if we could stick it out another twenty-four hours without rest. In the late afternoon, we emerged into a great burned track and the mountain rose majestically right in front of us. Banked to the top with snow, it looked like a great white cloud that had settled to earth for a moment's rest.

The open sky and the mountain acted like magic on my men. They began to talk and call to each other in laughing voices. I knew, then, that the strain was over; that they would not desert now. That night we camped in the shadow of the mountain, well out in the burned area, beside a pond of snow water. I slept for fifteen hours, utterly exhausted. In the late afternoon I shot a roe deer and that completed the contentment of our party. Four days at the Long White Mountain was sufficient. It was not important to climb to the summit for Younghusband had been there already. My object was to find what lay in that Korean wilderness over which he had looked thirty-three years before. We had a compass-line straight through the forest to the base and a rough map of the traverse.

I determined not to return by the way we had come but to strike through the wilderness to the headwaters of the Yalu River which could not be far to the west. It was a tough trip; just about what we had experienced on the way to the mountain. Dense forests, swamps, and drizzling rain. But the men pushed on with light hearts, laughing at difficulties and hard work, confident, at last, that my little compass knew the way.

Not far from the Yalu, I stumbled into the camp of eight Manchurian bandits, but talked my way out of a ticklish situation. The next day, we camped on the bank of the great river which at that point was only a rushing mountain torrent hardly a dozen yards across, and followed down its winding course to the first lumber camp on the edge of Korean civilization. There I sent back my men and floated downstream on a huge log raft, living in a little bark hut, sleeping in the sunshine, and watching the river widen and swell as it received the waters of a hundred tributaries. The trip was very restful after the weeks of strenuous work. I caught fish, shot ducks, and lived like a king. But a disreputable-looking person I was, with clothes in rags and shoes almost gone, when we reached Antung at the mouth of the Yalu River. On the train to Seoul I met Dr. Gale, an American missionary.

It seemed strange to talk English again for during all those months I had spoken nothing but Japanese. It was a coincidence, too, that Dr. Gale should be the last white man I had seen when the expedition began on the east coast, and the first one to meet me when it ended on the west side of Korea. He greeted me as one risen from the dead. Because I was long overdue some ambitious reporter had sent dispatches to America that I probably was lost in the northern wilderness and never would be heard of again.

On my part, I wanted all the news. He told me something of what had happened in recent weeks and then said he was going to America soon, via Europe, and asked what boat I planned to take across the Atlantic; perhaps we would be together. "There was," I said, "a great ship just about completed when I left New York. She was called the *Titanic*. I'd like to go on her if I can get a passage." He jumped as though from an electric shock. "What! You don't mean you haven't heard about her? Why she's at the bottom of the sea. I didn't think there was anyone on earth who didn't know about the *Titanic* disaster."

How could I? It had happened just after I went into the forests and my isolation was as complete as though I had been on Mars. A continent might have sunk beneath the sea and I should have been none the wiser.

Then in Seoul I had an amusing experience: I read my own obituary. The reporter had written a vivid story of my probable death and appended a neat little summary of my life and work. A cable to the Museum brought a joyous reply for they were seriously worried since I was weeks overdue. From Professor Osborn's message, it was evident that the results of my first independent land expedition pleased him enormously. A considerable area of unknown country had been explored and the Museum collections were richer by several hundred animals and birds, some of which were new to science.

With the Korean job behind me, and well done, I felt I had a sufficiently good excuse to wander for a while. Where to go? Peking first, of course, for China lay just across the Yellow Sea. After that I'd see whither the winds of destiny carried my willing feet.

When I arrived in Peking, the revolution which toppled the Manchu dynasty from the Dragon Throne, where it had sat for three hundred years, had barely ended. China was enduring the birth pangs of an infant republic which threw her into political chaos for four decades. The boy Emperor, Pu Yi, now the puppet of Japan in Manchuria, was a virtual prisoner in the northern part of the Forbidden City. Peking was fairly quiet, yet every day open carts carrying frightened men stripped to the waist and tightly bound drove through the city toward the "Heaven's Bridge," there to have the executioner lop off their heads like cabbages in a public display. It was just across the road from the "Temple of Heaven." It wasn't a pleasant sight, those head-lopping parties, but the people seemed to love them judging by the crowds that followed the tumbrils on their gruesome journey.

One execution I witnessed because of natural curiosity. Such a morbid strain exists in everyone, I suppose. Usually one such spectacle is more than enough. It was in my case. Eighteen men lost their heads that day. They kneeled in a long line and the executioner with his broad-ended sword started work as nonchalantly as though he were chopping trees. He made a mess of the first one, on purpose, and then began to argue as to who was going to pay for the job. With that small detail settled to his satisfaction, he went down the line in a thoroughly businesslike manner. One stroke, one head! Never will I forget one fellow twisting about and peeking up to see how soon his turn would come. That cured me from ever wanting to see another execution. I couldn't rid myself of the mental picture for weeks. I have seen many men killed since that time, but never when it could be avoided. An execution is about as attractive to me as a pest house!

I liked Peking from the moment I first walked through the Water Gate to the "Hotel of the Six Nations" where Manchu ladies with fantastic headdresses and painted faces sipped tea beside silk-gowned Mandarins. I liked the street cries, and the pigeons with whistles on their tails, and particularly I liked the walls. Those gray old walls, and their pavilioned corners, pregnant with romantic history, seemed almost to speak as I walked along their broad summits and peered through the crenelated battlements down to the shaded streets sixty feet below. The stations for the archers were plain to see as were the sockets for the Manchu banners set deep in solid stone. In the gate pavilions, the racks and pegs still showed where spears and shields and armor hung in orderly rows ready to be donned at a trumpet's blast by the defenders of the ancient capital.

The greatest wall on earth I went to see almost with reverence. I remember what a brilliant day it was, and how when I left the train and walked up the rocky pathway my eyes were steadily on the ground so that the Great Wall might appear in all its majesty suddenly, not bit by bit. When I looked up, at last, there it was, spreading its length like a slumbering gray serpent over the hills, into the valleys, and up the sides of precipitous peaks as far as my eyes could reach. No other sight on earth has ever stirred me as did the Great Wall of China. Not the Sphinx, nor the Taj Mahal by moonlight, nor the Pyramids, nor any of the Seven Wonders of the World. When I left Peking to go southward, I knew that some day I should return.

I have forgotten just why I decided to go to Australia, because it didn't attract me much, but anyway I took ship for Shanghai and Hong Kong. There a friend en route to Europe via Russia appeared in the lobby of the Hong Kong Hotel. "Why not come with me? What has Australia got that Russia doesn't have?" I didn't know, but was willing to find out, particularly as I had met a young Russian prince in Japan only the year before. He had been shooting in India and we became great friends. "If ever you come to Russia," etc., etc. So I sent him a cable and he said he'd meet me in Moscow.

The trans-Siberian train was like a ship, for the journey took nine days. Then, one afternoon, the spires of Moscow showed against a flaming sky and next morning I waked to see the sun glistening on a hundred cathedral domes, and to hear the chime of bells welcoming a new day. My Russian prince came before noon to take me to his wonderful place outside Moscow, where for two months I lived a kind of life that is now gone forever.

There were only two classes of people in Russia, then—the very poor and the very rich. One cannot imagine such luxury, such a total disregard for money, such pampering of every whim as characterized the nobility of Russia before the Revolution. It was "fine work if you could get it" to be a member of the ruling class, but equally hard on those who were not so fortunate. Having seen the play-day of the noble's life it is not difficult to understand why the pendulum swung to the opposite extreme as the clock ran down.

But don't think that I was so worried over the plight of the underdog that I did not enjoy what was offered me with such a lavish hand! We rode and shot, danced and played, drank and ate—particularly did we eat. Food such as I have never dreamed existed, with vodka and caviar served in huge blocks of ice, seemed always on the table. Fortunately for me, at that time it was considered plebeian to speak Russian except to servants. In social intercourse, English, German, and French were always used. French, perhaps, most of all, but virtually everyone I met knew English. Of course, I fell violently in love (or thought I did) with a beautiful Russian girl who reciprocated my affection as a matter of hospitality and promptly forgot me when I left.

Eventually, my conscience became a little bothersome, and to stop its annoying pricks I turned homeward. After two months of wandering in Finland, Sweden, Norway, and Denmark, I sailed for New York on a Danish ship, and got home with all my money spent, as usual.

By a pleasant coincidence, the day I returned to New York Sir Francis Younghusband was at the American Museum to lecture on his Lhasa expedition. I had the satisfaction of telling the great explorer what lay in the gloomy forests of Korea over which he had looked thirty-three years before. It had been his book and encouragement that had sent me on the expedition. From that time on, he took a personal interest in my career as an Asiatic explorer and we became great friends. I never missed seeing him when in London and he came several times to New York. It was a real sorrow to me when I had news of his death in 1942.

Chapter 12

A Harem on the Rocks

For once, when I reached the Museum, I did not have plans for a new expedition. My work at Columbia University had been on a hit or miss basis; mostly miss, for I hadn't been able to stay put long enough to do consecutive resident work. A study of the redis-covered California gray whale would make an excellent thesis, so I plunged into it like a swimmer taking a dive in pleasant waters. All day and most of the night I worked and in six months got it ready for my degree in June. Only just did I get under the wire, for John Borden of Chicago had built a yacht, the *Adventuress,* and planned an expedition to the Arctic to catch a bowhead whale, which the Museum badly needed to complete its collections of big Cetaceans.

John was a keen sportsman, and a fine fellow, who had recently inherited a large fortune and was having a grand time spending it in the way that gave him most fun. Polo and big game hunting had begun to pall, so he conceived the idea of building a yacht and harpooning whales in the good old-fashioned way. The plan was sound but its execution not so good. The boat was lovely but completely unfitted to go into the ice; moreover, John invited some friends who were less suited for the expedition than the yacht. That trip taught me that yachting parties and serious natural history work do not mix. Since then I have been invited to go on many such trips, and when I was director it seemed that something of the sort was offered to the Museum every month. The trouble with that kind of party is that the guests get bored while naturalists are working. Templeton Crocker of San Francisco was the only exception I ever met. He was deeply interested in natural history and his trips were for naturalists only; there were no other guests. He completed two highly successful expeditions for the Museum on his yacht *Zaca*.

The *Adventuress* cruise to Alaska turned out to be nothing but a big game hunt. Under ordinary conditions it would have been delightful, but I had come to do a job and was stymied by circumstances. On the Alaska peninsula we shot caribou, emperor geese, ptarmigan, and ducks; hunted mountain goats along the precipitous shores of the beautiful fjords and killed great brown bears under the shadow of Pavlof volcano and on Kodiak Island.

It was there that I had a strange experience. Salmon were running into the creeks and rivers by the millions to spawn and die, and the bears, who love fish, were having a field day. Walking up a partly dry stream bed, I came to a great log jam. I had almost reached the top of the barrier when the head and neck of a bear appeared above another jam about a hundred yards away. An enormous head it was—like a full moon. I was shooting my little 6.5 Mannlicher and lined the sights just under the snout. At the crash of the rifle, the head disappeared but in a second it bobbed up again twenty feet to the left.

"What the devil is the matter with me?" I thought. "I couldn't have missed at that range."

For the next shot I rested my rifle against a stub and squeezed the trigger. Down went the head. Climbing to the top of the jam, I could see my bear running away. A quick shot rolled him over but he was up again in a second. The next bullet put him down for good. When I reached the animal I found an enormous female, very fat. She must have weighed close to thirteen hundred pounds. I looked her over carefully trying to see if my first two bullets had touched the neck. Not a sign of a wound anywhere. Feeling very disgusted at my bad shooting, I straightened up wondering how I was going to skin the beast. To my intense surprise there lay a second bear at the foot of the log jam and twenty feet away still another, both stone dead. It was a whole family—a mother with two full-grown cubs almost as big as herself. The Alaskan brown bear is the largest living carnivore. A big one will weigh about fifteen hundred pounds. In the spring after their long winter's sleep they are ravenously hungry and will come at a man like a shot. At the end of the summer, when rolling in fat, it is too much effort to attack unless cornered or provoked. But it is wise to remember they'll not take anything from anybody.

I did one job on the *Adventuress* cruise that was important and interesting. Dr. Hugh M. Smith, Commissioner of Fisheries, had asked me to take motion pictures of the seal herd on St. Paul's Island in the Bering Sea for the U.S. government. No motion picture camera had ever been allowed on the island. Why, I don't know. The Bureau would give me all facilities and I could have a print of the negatives for lecture purposes. It was a real opportunity.

John Borden dropped me at St. Paul's while the yacht went southward and I was to come out on a revenue cutter which would make its last trip of the season in about three weeks. From that time until spring the community on the island would have touch with the outside world only by wireless.

During the long winter the black rocks of the Pribilofs lie deserted, stark and cold. But in the early spring the old bull seals come up from the south. Great hulking fellows they are, rolling in fat, and filled with rutting wrath. Each one selects for himself a private station on the shore, destined to be the site of his prospective harem. The best localities are those nearest the water where the females will come to land.

But just because an old bull happens to get there first does not mean that he is left quietly in possession. Far from it. He has to defend his claim against all comers. The choice sites for domestic establishments are soon gone and later arrivals either have to fight for what they want or take the back lots away from the beach. For days the shore is a roaring, bloody battlefield. Rearing on their hind flippers, the bulls throw themselves at each other, slashing viciously with their long canine teeth, shoving, pounding, tearing. Perhaps the original occupant is driven out. Then the newcomer must hold his claim in daily battles. All this is preliminary to the real show. That begins when the first soft-eyed female pokes her sleek head out of the water in the white surf line.

"The girls are here. The girls are here," roar the bulls. The beach seethes with excited swains. Modestly, the little seal ladies swim toward shore, play coyly about in the surf, draw off to deeper water, tantalizing the impatient gentlemen bouncing up and down on the rocks of their hard-won homes.

When the females finally decide to land, the fury begins. Each bull rushes for the nearest one, trying to entice her to his home site. Pushing and shoving, fighting off other aspirants for the lady's flipper, he finally drives or beguiles her to his station. Once there he has little time for love-making for other females are constantly arriving. The bull seal believes in quantity rather than quality. He never seems satisfied. His harem consists of from five to sixty and they certainly keep him busy. Each time he makes a dash to the water's edge on another amorous excursion some neighbor tries to steal a wife or two. Perhaps one of the females has been attracted by the beautiful curling mustache or deep bass voice of a bull just a few yards away. She gives him the high sign and makes a dash for freedom while friend husband's back is turned. She may gain a new home. More often she is unceremoniously hauled back and told not to try *that* again.

The rushing season ends when the females have all arrived. Then the beach is turned into a maternity ward. Almost every lady seal gives birth to a little black squalling baby within a week of her arrival. As a matter of fact, that is what she has come for. The conceited bulls think it is their masculine charms that bring the females shoreward. They are dead wrong. It is only mother instinct. She wants her baby born in the old homestead where she herself first saw the light. She is a pretty good mother, too. Hitler would give her an Iron Cross. "One each year," is her motto and she does her duty conscientiously every spring.

But the old bull has a hard season. All through the summer he neither eats nor sleeps. It is just one long debauch of lovemaking and fighting and guarding his harem against unscrupulous invaders. By September he is a wreck of his former self. All his fat has disappeared for that is what he has been living on by absorption during the summer. His bones protrude, his hide is torn and scarred, he is weary unto death. Blessed sleep is what he needs. Forsaking his harem, he waddles back into the long grass, far away from the beach, there to stretch out in the warm sun. He will sleep for three weeks without waking if left undisturbed.

The little black babies don't have such an easy time of it during the summer. Many of them are crushed by their fighting fathers. Sometimes they wander a bit too far and lose their mothers. I often watched a hungry pup scramble up to a dozing female and settle down for a drink of milk. She might turn, sniff disgustedly at the youngster, and give him a slap with her flipper. None of that for her. She wasn't any public fountain. Her own baby needed all the milk she could furnish and she wasn't giving any away to stray guttersnipes. I could understand what she said just as though she spoke my own language. If the baby was overpersistent, she would get really angry, cuff him hard, and send him away squalling. Usually his own mother soon found him. I frequently saw a female seal scrambling about wild-eyed exactly like any human mother looking for her child.

"Where has Willie gone? I left him right here when I went out to fish. Willie! Has anyone seen my Willie?"

Then such cooing and gurgling when the lost had been found! She would gather the whimpering black pup to her breast, stretch out on the rocks, and let him feed to his stomach's content. The pups all looked alike to me, but every mother knew her own child right enough. Doubtless, it was by odor rather than sight.

The pups were courageous little things. If I picked one up he spat and swore like a trouper, doing his best to bite with his tiny needle-like teeth. I never knew that seals had to be taught to swim. I thought they knew how to paddle instinctively, like a duck. Not so. I used to watch the mothers giving swimming lessons in the tide pools. The babies were afraid of water. They didn't want to go in at all. Slaps and vigorous cuffings were required before they would even get their flippers wet. Sometimes the mothers had to throw them bodily into the pools. But once in they learned the motions quickly enough.

Although the old bulls never left their harems, the females and the bachelors went out daily to fish. Bachelors are those seals that have not yet reached adulthood and the dignity of a harem. They are the one- and two-year-old males, and theoretically are the only ones to kill for fur. The skin of an old bull is valueless. It is too thick and heavy and too scarred by fighting.

I used to lie behind the rocks watching the seals by the hour. Their eyesight is so poor that unless I showed myself against the skyline they seldom took alarm. When I walked boldly out on the beach there was a general and hurried exodus to the water. I arrived at one rookery where there were six thousand seals, in the midst of a terrific gale. The surf pounded in tremendous breakers upon the rocks. For hours I lay concealed, shooting bits of intimate seal life with the movie camera. Then I walked out upon the shore line. Like a church congregation standing to sing, the six thousand seals rose as one. For a few moments they gazed at me and then broke for the beach. In a living flood they poured over each other down the rocks and into the water. Riding the breakers like surf boats they floated in on the waves and out again while I ground off hundreds of feet of film. What a picture it was! Had I remained all summer I might not have got another like it.

When the revenue cutter arrived I left St. Paul's with regret. As the ship moved out of the bay every resident was gathered on the shore in a sad-faced group. It was their last contact with the outside world for many months. Day after day the ice of the Bering Sea and winter gales would pound against their rock-bound home. They had the wireless, to be sure, but no matter what happened aid could not reach them. The cemetery on the hill slope opposite the station tells its own story of those wanderers who have died a lonely death on the cheerless wastes of this bleak island.

Chapter 13

The Blue Tiger

A few weeks after the First World War began in August 1914, I remember talking with a prominent New York banker.

"How long," I asked, "do you think the war will last?"

"Well," he said, "from an economic standpoint, I don't see how it can go on more than six months. At the rate they're spending money there just isn't enough cash or credit to carry Germany beyond that. After all, you know they've got to have one or the other to get the things with which to wage the war."

That gave me a certain amount of comfort because bankers are supposed to know about such things and one trusted them. Be it said, moreover, that my particular banker was far from standing alone in his idea. People in those days couldn't conceive of a world wherein a million dollars an hour could be spent in order to kill other people, with all those dollars going out in gun flashes and nothing coming in by way of international trade.

Like most Americans at that time, I felt it was a European show and that we had better keep strictly out of the mess. In September, I think it was, I had luncheon with ex-President Theodore Roosevelt at the Century Club and remember how bitterly he denounced the Germans for going through Belgium. "Mark my words," he said, "we'll be in it ourselves before long. In decency we can't stay out. Wilson wants neutrality when everything sacred in international relations has been violated. He asks us to think and

act neutral! Neutrality! Poppycock! He's got a chocolate eclair
for a backbone!"

I didn't, nevertheless, agree with the Colonel that we should
leap to arms in Europe's quarrel. But it was obvious that a person
with any sort of convictions couldn't think and act neutral. That
was just plain silly. You were either for the Germans or against
them. It was amazing, too, how many people were for them in the
beginning. It was only when they began the submarine warfare
and sank a lot of our ships that German sympathizers disappeared
from one's personal acquaintances. Then, of course, the *Lusitania*
gave it the *coup de grâce*.

I was somewhat confused in those early days of the war, for
it was just at that time that I married. So many other things
were of direct personal importance that the war seemed almost
academic.

I was living in Lawrence Park, Bronxville, in 1915, and remember
going to New York the morning of May 8 with Caspar Whitney,
the sports arbiter, explorer, and writer. We were talking about an
expedition to eastern Tibet which I had in mind, and were blissfully
unconscious that anything unusual had happened. At the station
an excited group of commuters surged about the newsstand where
the papers flaunted great black headlines announcing the sinking of
the Cunard liner *Lusitania* by a German submarine. One hundred
and fourteen Americans had lost their lives with the 1,153 other
passengers!

Almost everyone expected, of course, that Wilson would go
before Congress immediately and ask for a declaration of war on
Germany. The feeling was one of tense waiting, with the public
willing to give the president a chance. But all that happened was
a series of note exchanges which developed into a diplomatic
debate.

A wave of disgust swept the country. Had Wilson declared war
at once my own life would have been radically changed for I should
have volunteered within a week. Ted and Kermit Roosevelt and
several of my other intimate friends were already helping fight

England's battle. I wasn't ready to do that, but *our* battle was a different story. I was patriotic but not altruistic. That, I think, was the attitude of most Americans at the time of the *Lusitania* sinking. Having settled this with my conscience, I turned to matters of personal importance.

I had decided to abandon the study of whales for land exploration. Always that had been what I wanted to do. After my first trip to the Orient, a paper by Professor Henry Fairfield Osborn, in which he expounded the theory that Asia was the home of primitive man and the mother of much of the animal life of Europe and America, was always in my mind. To explore Central Asia and test his theory had gradually become almost an obsession with me. Then in February 1915 Dr. W. D. Matthew published his brilliant paper, *Climate and Evolution.* It was an elaboration of Osborn's thesis and fired me with enthusiasm. I wanted to begin at once. But Central Asia was a forbidding place and virtually unknown. One couldn't go there just like that. First, I had to know where to go, learn the customs, methods of travel, and the language. I could speak Japanese fluently and some Korean but Chinese would be most useful of all.

I thought it all out and then went to Professor Osborn. I wanted, I told him, to make a really scientific attack on Asia but the first expedition would work along the margin of the Central Asian plateau. It would be purely zoological, bring to the Museum valuable collections of mammals, birds, and reptiles, and incidently teach me some of the things I needed to know before I was ready to go at the heart of the continent in a big way.

Professor Osborn was a man with great vision and a believer in young men. He was an explorer at heart, but because his life had been cast in academic halls, he did his exploration by proxy. I can think of no man in the world who has been responsible for so many expeditions. Since my idea was eventually to test his own theory which had been evolved after years of research and thought, he was more than sympathetic. But there was the question of money, as always. I wanted to explore Yunnan, the mountainous province of southeastern China, which margined the Tibetan plateau. The expedition would cost fifteen thousand dollars and I agreed to raise half of it among my friends if the Museum would provide the remainder.

I don't remember where all the money came from. I think Sidney M. Colgate, James B. Ford, Charles L. Bernheimer, George S. Bowdoin, and Henry C. Frick gave most of it. Anyway I got it, and by March 1916 we sailed on the Japanese ship *Tenyo Maru.* Edmund Heller was the other scientist of the expedition. Heller had accompanied Colonel Theodore Roosevelt on his African trip after he left the White House and was an excellent small mammal collector, although hardly as successful a field companion.

The expedition had been largely publicized because one of our objectives was to hunt the so-called "blue tiger" in the jungle of South China. The phrase caught on, of course, and the newspapers had a lot of fun with "blue tigers," "blind tigers," and the like. It made everyone curious and we went off to the accompaniment of popping flashlight bulbs and screaming headlines. They continued at Yokohama when the ship pulled into the long wharf and a group of Japanese approached. Bowing in unison one said, "We are report for leading Japanese newspapers. We wish to know all thing about Chinese animal."

The blue tiger had been discovered by Harry Caldwell, a shooting missionary who had made quite a reputation for himself in China. With a Bible in one hand and a rifle in the other, he pursued his evangelical work among the Chinese, ridding their villages of man-eating tigers, the while he poured into their ears the Eternal Truth. Harry is a regular fellow. Bursting with enthusiasm, interested in everything under the sun, but particularly natural history, he has done much real good among the people of his district.

Twice Harry had seen the tiger at close range and the natives knew it well. The beast had become a man-eater and was credited with many victims. Caldwell described it as striped with black on a blue-gray ground. Doubtless it was a partially melanistic phase of the ordinary yellow tiger. Melanism, the opposite of albinism, occurs very frequently in some animals but is rare in others. Black squirrels, of course, are common and so are black leopards, but no dark-colored tigers had previously been reported.

The animal ranged in a district about Futsing not far from Foochow. The method of hunting was the same I had followed in Korea. Hard traveling from village to village wherever the beast was reported by the natives. The heat was terrible. Wet soggy air, blazing sun, fast traveling, hard climbing. For a month we kept on his trail but he was as elusive as a will-o'-the-wisp: the personification of the "Great Invisible."

Once we missed him by sheer bad luck. Natives had seen him go into a deep ravine early in the morning. It was a wild place seemingly cut out of the mountainside with two strokes of a mighty ax and was choked with a tangle of thorny vines and sword grass. Impenetrable as a wall of barbed wire, the only entrance was by the tiger tunnels which drove their twisting way through the murderous growth far in toward its gloomy heart.

Caldwell and I staked out a she-goat with her kid in a small clearing. We concealed ourselves in the grass on a slope thirty feet away. The kid kept up a continual bawling for its mother. The lair shimmered in the breathless heat, silent save for the echoes of the bleating goats. Crouched behind the screen of bushes for three long hours, we sat in the patchwork shade, motionless, dripping with perspiration, hardly breathing, and watched the shadows steal slowly down the ravine, pass over us, and reach a lone palm tree on the opposite hillside. By that I knew it was six o'clock. Suddenly at the left, and just below us, there came the faint crunching sound of a loose stone shifting under a heavy weight, then a rustling in the grass. Instantly the captive goat gave a shrill bleat of terror and tugged frantically at the rope which held it to the tree. Harry put his mouth close to my ear and whispered, "Get ready. He's coming."

The blue tiger was there, not twenty feet away, taking a last survey before he dashed into the open to leap upon the goat. I sat waiting, every nerve tense, the butt of the rifle nested against my cheek. Suddenly pandemonium broke loose. Over the hill opposite to us came a party of woodcutters. Blowing horns and banging tin pans, they streamed down the slope across the bottom of the ravine and up the other side. They were taking a short cut home and the noise was to drive away the tigers. Our chance was gone.

In spite of the fact that my companion was a missionary I did some really artistic cursing. Had those damned Chinese arrived ten minutes later the blue tiger would have been lying dead in front of us. Instead, he had quietly withdrawn into the depths of his lair. We found the great pug marks where he had crouched at the entrance to the tunnel ready to charge into the open.

It was rotten luck but the old men of the village wagged their heads and said, "You never can kill him. He is not a proper tiger. It is an evil spirit." It did seem that the beast was a phantom possessed of some occult power to evade the waiting death that had stalked his heels for more than a month. A few days later a Chinese found the blue tiger asleep under a rice bank and dashed frantically to our camp only to find that we had left for another village half an hour earlier. Again the tiger pushed open the door of a house at daybreak, stole a dog from the "heaven's well," and devoured it on a hillside near the village.

After we left, Caldwell had another go at the blue tiger but never saw him. Others have hunted him since, but at the time of my last report he was still at large. His death toll had reached nearly a hundred people.

When the tiger episode was ended we hied ourselves to Hong Kong for food and final equipment. There at "Lane and Crawfords" I learned an important fact. My forty or fifty boxes were coming into being with aggravating slowness. I said to the manager, "Can't you hurry those carpenters?"

"No," he replied. "After you have lived in China for a few years you may possibly learn that the only way to hurry a Chinese is to get more Chinese. We'll put ten additional carpenters on the job tomorrow."

We came to Yunnan via the then little known island of Hainan and French Indo-China. At the port of Haiphong a French train took us to Hanoi, the capital, and thence through the jungle plains into the mountains of China's most beautiful province to Yunnan-fu, now called Kunming. Nine thousand feet high, near a lovely lake and blessed with the sort of climate that makes old men feel young, Yunnan-fu was a place to write home about. It was the terminus of the now famous Burma Road. Our evening clothes had been packed in moth balls but they were airing in the sun the next morning. Dinner parties, dances, and all the activities of a summer resort were the order of the day. That's what it was, a summer resort; for the wives of French officials who could not, or would not, stand the heat of the coastal plains, flocked to play in the sparkling sunshine of China's Switzerland.

There was a delay of some days in Yunnan-fu because a young and very attractive American who was to handle my financial affairs was temporarily incommunicado.

"He isn't home," they told me at his office. "He's away. He'll be back sometime. We don't know just when. No, you can't reach him. He's gone. He isn't here."

I smelled a mouse because his head office in Hong Kong, which had given me a letter of credit, said he *was* there and would be all summer. It took very little snooping to discover that everyone in Yunnan-fu knew his whereabouts, which weren't far away. To be exact, he was sitting in a houseboat in the middle of the lake with field glasses glued to his eyes. All because of the beautiful wife of a French banker of Hanoi. She had come to Yunnan-fu to escape the heat and a dull husband. My "financial agent" certainly wasn't dull, nor had he been slow in making hay while the sun shone. The result was that reports of the goings-on had reached the banker and in due time he arrived in Yunnan-fu with the announced intention of satisfying his honor at the point of a duel. Word of his coming had been telegraphed ahead, and my agent's friends parked him out on the lake in a houseboat, with the lady and a plentiful supply of food and champagne. Moreover, a vigilante committee guaranteed that no other craft would be available for the outraged husband. In some strange French way "go betweens" satisfied his honor and he departed for Hanoi, there to divorce his wife. When the banker was safely on the train, the houseboat was wigwagged and my agent gave orders for "anchors aweigh." A temporary bar had been hastily arranged at the dock by the welcoming committee, and most of Yunnan-fu's foreign residents greeted the returning voyagers with cocktails and champagne.

Such was Yunnan-fu and many other parts of the Orient in those gay careless days. One laughed at life and death, but always one laughed. Weep, and you wept alone. It was fun if you were willing to pay the price but there was a price to pay. The tragedies came if a man, more usually a woman, thought that when the dance ended somehow, some way, one could avoid the bill.

Yunnan Province did not rate an American representative so I went to the French consul, M. Wilden, for help with the Yunnan Foreign Office. Never was there a more charming man than Wilden. Later he became consul general in Shanghai and then French minister to China. It was a great loss to countless friends when he died. Wilden arranged an interview with the commissioner of foreign affairs, a fat, jolly little man who received us at half past ten with an elaborate "collation" and quantities of the vilest sweet champagne I have ever had to taste. I had an open letter of introduction from Colonel Theodore Roosevelt which enormously impressed the commissioner. He remarked with a naive smile: "After Colonel Roosevelt finished his term as president of the United States we asked him if he would not come here as governor of Yunnan Province but he replied that he had other commitments."

I could just see the Colonel in the midst of Chinese politics!

Chapter 14

Yunnan Adventures

Always I have been a believer in conservation. I conserve old hats, old shoes, old pants, and particularly old slippers, to the disgust, I may say, of my wife. But when we left Yunnan-fu for the wild hinterland on what is now called the Burma Road and I saw the nth degree of conservation by the mule drivers, it made me feel like a piker. An ingenious little basket fastened to the crupper of each plodding mule hung just where it would do the most good. Not one precious fragment of fertilizer got past the basket, and at the end of the caravan march the *mafus* counted their gains according to what had gone in and come out. This they sold when we reached a town of sufficient importance where merchants had seats on the manure exchange. If there was a depression in the market, due to the passing of many caravans, the yield was carefully hoarded until the price had risen where it could be unloaded at a profit of more than one hundred per cent. Not ninety-eight, or ninety-nine, but one hundred per cent clear profit was demanded from the hard-boiled exchange brokers.

The Chinese proverb that a "road is good for ten years and bad for ten thousand" applied perfectly to what passed for a highway in 1916 when we traveled over its entire length from Yunnan-fu to the Irrawaddy River. It was paved with huge blocks, broken and irregular, standing up at impossible angles. Where they were still in place they had been worn to such glasslike smoothness that one slipped and slid as though walking on ice. As a result, every caravan avoided the paving whenever it could find a path, and sometimes dozens of deeply cut trails spread over the hills beside the road, only coming back to it when a ravine or impassable slope made other routes impossible. It was a highway which wound over mountain peaks, down into valleys, across rivers on swinging rope bridges and finally reached the jungle-filled plains of Burma and the Irrawaddy River at Lashio or Bahmo. How it could have been converted into a route for motor cars, God and the Chinese only know. Manpower, the one thing that China has the most of—men by the hundreds of thousands, digging like moles into the slopes, was what did it.

For centuries the road has been one of the main trade arteries through the province, bringing the goods of peace from India just as in the last four years it has brought the goods of war. Marco Polo traveled over it on his trips for Kublai Khan. In his book, he describes the beautiful lake of Tali-fu and tells of a city where ducks were split, salted, and dried as in no other place in China. Sure enough, there they were, hanging in the market place, when we rode through the town seven hundred years after he had passed that way.

And on the plains near Yung-Chang where we camped, one of Kublai Khan's generals, Nestardin, he says, fought a great battle with the king of Burma in 1272. There, for the first time, the Mongols saw elephants. Their horses were terrified and bolted into the woods but Nestardin knew that the strange creatures must be made of flesh and blood. So he ordered his men to tie their ponies to the trees and advance on foot, shooting clouds of arrows at the elephants. Bristling like pin cushions, trumpeting with pain, the great beasts dashed away, throwing the king's troops into confusion and giving victory to the Mongols. Over this same road, Nestardin brought back dozens of elephants to the court of the great Khan where they were used in his armies ever after.

Less than a hundred years ago the road was slippery with blood when the Chinese turned against the Mohammedans who had come over it up from the south and slaughtered them like sheep. But when we were there, it was a road of peace and the ordinary affairs of life. Pigs, chickens, horses, and cows lived in happy communion with the human inmates of the villages. The pigs, especially, were treated with the utmost favor. On the doorsteps, children played with the swine, patting and pounding them, and sometimes a mother even brought her baby to be nursed by a sow with the family of piglets.

While we were on the Burma Road we camped in temples, some of them half ruined during the Mohammedan rebellion. The priests were always glad to have us there for it meant money in their pockets and surpassing interest for their dull days.

On the ninth march there was a bandit scare when we were in a bad place, but we got ourselves to the summit of a pass where we could look down upon a caravan being robbed. Right below us the brigands were ripping open packages and scattering their contents right and left. There were forty of them. I could easily have killed half a dozen, but it wasn't our show. My motto, while traveling in the Orient, has always been to keep my hands out of other people's business, particularly where bandits are concerned. It has saved a lot of trouble.

At last the brigands found what they wanted, several packages of jade, and disappeared into the mountains. We learned, later, that they had been following this caravan for several days, waiting for the proper moment to attack. It belonged to a rich mandarin and the bandits knew exactly what he had among his possessions. So far as my expeditions were concerned, I always let it be known that we carried no money or anything bandits could use except rifles, and that they were ready to be used.

From Tali-fu, almost in the center of Yunnan, we left the Burma Road and traveled far into the north even into the borders of Tibet where the wild aboriginal tribes lived in secluded mountain valleys, still using cross-bows and poisoned arrows. Wherever we went we searched for animals—big mammals, small mammals, rats and mice and shrews, and birds, frogs, reptiles, and fish. For this was a zoological expedition pure and simple for the purpose of bringing back to the American Museum a cross section of the animal life of this little known region.

From Li-chiang, a frontier village with a curious mixture of Chinese, Tibetans, Mosos, and half a dozen other tribes, we rode straight to the Snow Mountain which rises twenty thousand feet above the sea. Below the summit in a grassy meadow beside a stream of green snow water, our tents were pitched just at the edge of the spruce forest. Then we climbed to the grassy slope above the timber line to set a hundred traps in the runways of meadow voles and under logs and stumps in the forest. Our Moso hunters, with their dogs, slept in their ragged clothes, without a blanket or shelter of any kind, beside a huge rock.

It was, I think, the most beautiful camp I have had anywhere in the world. Above our heads, silhouetted against the vivid blue of a cloudless sky, towered the great Snow Mountain, its jagged peaks crowned with gold when the light of the waning sun touched their summits. As we sat at dinner that night about the fire, and the light faded, we could see the somber mass of the forest losing itself in the darkness and felt the unseen presence of the mighty peaks standing guard over our mountain home.

We were awakened before daylight by our Boy's long drawn call to the hunters, "*L-a-o-u H-o, L-a-o-u H-o.*" It was a cold, gray morning with dense clouds weaving in and out among the peaks, but nevertheless I decided to try for goral, the Asiatic counterpart of the European chamois. It would be a new animal to add to my list of big game. Two of the men took the pack of mongrel dogs around the base of a high rock shoulder, sparsely covered with scrub spruce, while I went up the opposite slope accompanied by the other two. In half an hour the hounds began to yelp and we heard them coming around the summit of the ridge in our direction. The Moso hunters made frantic signs for me to hurry up the steep slope but in the thin air at fourteen thousand feet my heart was pounding like a trip hammer and I could move at only a slow walk. Suddenly the dogs appeared on the side of a cliff just behind a bounding gray form. The mist closed in and we lost both hounds and animal, but ten minutes later a blessed gust of wind swept the fog away and the goral was indistinctly visible with its back to the rock, facing the pack. The big red leader now and then rushed in for a nip at the animal's throat, but was kept at bay by its vicious lunges and sharp horns.

It was more than two hundred yards away, but the cloud was drifting in again and I dropped down for a shot. The first two bullets were low, but for the third I got a dead rest over a stone and at the roar of the little Mannlicher the goral threw itself into the air, whirling over and over onto the rocks below. The hunters, mad with excitement, dashed up the hill and down into the stream bed. When I arrived, the goral lay on a grassy ledge beside the water. Then began the ceremonies which must be performed at the death of the first animal in every hunt. One of the Mosos broke off a branch, placed the goral upon it, and at the first cut in the skin chanted a prayer. Laying several leaves one upon the other, he sliced off the tip of the heart, wrapped it carefully, and placed it in a nearby tree as an offering to the God of the Hunt.

All of our Moso hunters, except one, were armed with cross-bows. They were remarkably good shots and at a distance of a hundred feet could put an arrow into a six-inch circle four times out of five. The one man who did not carry a cross-bow boasted of a most extraordinary gun. Its barrel was more than six feet long with the stock curved like a golf stick. A powder fuse projected from a hole in the side of the barrel and just behind it was a forked spring. At his waist, the man carried a long coil of rope, the slowly burning end of which was caught in the crotched spring. When he wished to shoot, the native placed the butt of the weapon against his cheek, pressed the spring so that the burning rope's end touched the powder fuse, and off went the gun. I tried it once, but never again; my face was sore for a week!

The first goral hunt was a pattern of the many hunts we had for serow, goral, muntjac, and other strange animals on the Snow Mountain and in the wild country along the Tibetan frontier. They were interrupted, however, for more than a month, by a severe infection in the palm of my hand from which I almost died. We had gone down to a temple at the base of the mountain when my hand suddenly began to swell. Twenty days' ride from a doctor, there was nothing to do but treat it myself with steaming cloths, during intervals of delirium. My arm swelled to twice its size, but eventually the poison subsided and I was well again, though my hand was useless for many weeks.

After leaving the Snow Mountain village, we pitched our tents on the banks of the "White Water" and later on the "Black Water" near a deep canyon of the mighty Yangtze where no man had ever been.

I remember, particularly, the day when we crossed a pass sixteen thousand feet high into the Yangtze drainage basin. In a few hours we climbed out of the warm October sun into the dead of winter. Up through a larch forest into the higher belt of dwarf bamboo, beyond the uttermost timber line of rhododendrons, to the summit of the pass, bare and bleak and frozen. A bitter wind swirled about our tents. It was too cold to sleep. All night we shivered about a tiny fire for there was little wood. Three of the ponies died from cold and the effects of the altitude. All of us suffered severely and it was a miserable party that descended next morning into the golden sunshine of the October we had left so suddenly.

Every mountain range brought us into new valleys occupied by strange aboriginal people. There are thirty distinct tribes in Yunnan, the remnants of the original inhabitants of China. Just as the white men pushed the American Indians westward, so did the Chinese drive the aborigines south and west centuries ago. Now they are concentrated in the mountains of Yunnan. One of them, the Lolos, never have been subdued by the Chinese. They still occupy a territory called Lolo Land in Western China where no Chinese is allowed to pass.

Small bands of Lolos have wandered from the forbidden country and settled in Yunnan. After crossing the pass, we came down to a Lolo village hidden away deep in a secluded valley. Fine, tall fellows they were, with long heads, high-bridged noses, and thin lips and faces almost Caucasian in type. They never had seen a white man and at first were frightened. Tobacco and small presents soon made them realize that we were friends.

Everything about us was interesting to them. Cameras, watches, and the like were too far beyond their comprehension to be impressive, but the field glasses seemed a miracle; also my high-powered rifle and revolver were tools of a god, for their own guns were primitive matchlocks with a range of only thirty yards. In photographs, they could not recognize themselves. It was only by pointing to some special article of dress and indicating it in the photograph that they could be made to understand.

Passing through the Moso country, up to the frontier of Tibet, we went, finding new mammals and birds, new plants, new tribes, and unmapped trails. Then across the Yangtze into the Mekong gorge and southward to the steaming tropics of the Burma border. Thick palm jungle instead of snow-capped peaks; leopard, tiger, sambur, and monkeys; peacocks, jungle fowl, and half a dozen other pheasants. A country as different from that we had left as Cuba is from Alaska.

The Salween River valley lay between us and Bhamo. I wanted to go there but hesitated, wondering whether it was worth the risk. A ghastly place it is, deserted of all human life, given over to peacocks, leopards, and wild red dogs. Even the aboriginal Lisos dared not face the malignant malaria which makes the valley a fever-stricken hell. It was here that General Stillwell's Chinese troops fought the Japanese only a few short months ago. In a copy of *Time* magazine, December 7, 1942, I read such a vivid description of the Salween River valley that I quote it here in full:

> *Dying weather.* Mountains, mottled green, yellow, red and gray tower thousands of feet into the air, drop precipitously into the emerald green Salween, called by the natives *Wu-ti-Ho*, the "River without a Bottom." In the jungles with the Chinese were leopards and tigers; pythons that swallowed whole live hogs; monkeys that stole soldiers' food, wolves that

howled at night and tried to steal dead soldiers. In the river, said the natives, were little fish with hides thicker than leather; bigger leather-skinned fish whose mouths opened and shut like folding doors. Some of the natives, ceremoniously neutral, stalked the Japanese with poisoned arrows; some hunted the heads of unwary Chinese.

But worst of all was the *ta-pai-tzu* (malaria). This was the worst malaria spot in the world. The deadly mosquitoes infested the gorge. Exhausted, underfed and ragged the soldiers had neither mosquito nets for protection nor quinine to combat the fever. Casualties from malaria were higher than from combat. Apparently well men, trudging along the mountain passes, would suddenly flush, complain of the fire in their heads, then die. It was months before adequate doses of quinine reached them.

The Japanese were better off on their side of the gorge. They had the southern end of the Burma Road, over which they could transport medicine and materiel, move their men back to base hospitals. For the Chinese, the section of the Burma Road which they held, winding on north to Chungking, was a broken impassable trail. They themselves had destroyed it to forestall any further Jap advance.

As I read of the present war, I realize that no unknown corners of the world remain. What were the ends of the earth twenty-five years ago appear suddenly in the headlines of every newspaper. The jungles of New Guinea, Borneo, and the Salween valley were mysterious and unknown. Today, they are the battlefronts of the world. Truly the romance and adventure of exploration are gone forever!

I decided to go to the Salween even though we were flirting with death, for other collectors had avoided it like a plague spot and I knew that whatever mammals we obtained would be new to science. There were gloves and mosquito nets for every man, and I was prepared to give deep injections of quinine, which is the only way successfully to battle the fever. So we went and remained ten days. Our reward was a fine collection of mammals and I was the only one who contracted fever. The mosquitoes bit me while I was waiting for peacocks on the river's edge, for I had lost my gloves.

That night I was shot full of quinine and the next day we climbed gratefully five thousand feet out of the poison valley to the healthy ridges where we could look down upon the river winding like a thin green line below. For three days I shook with chills and burned as though there was fire in my blood, but the blessed quinine brought me through and never has the fever returned.

A few weeks later, while hunting the black gibbons of Ho-mu-shu, I had a terrifying experience. We were camped on a sharp mountain spine, less than a hundred yards wide. The sides fell away for thousands of feet in steep forest-clad slopes and, as far as our eyes could reach, wave after wave of mountains rolled outward in a great sea of green. The trees were immense, spreading giants with interlaced branches that formed a solid roof one hundred and fifty feet above the soft moss carpet underneath. Every trunk was clothed in a smothering mass of vines and ferns and parasitic plants; from the lower branches thousands of rope-like creepers swayed back and forth with every breath of wind. The canopy was so dense and close that even at noon there was little more than a somber twilight beneath the trees.

While the tents were being pitched, the forest suddenly resounded with a wild "*hu-wa, hu-wa, hu-wa!*" We grabbed our shotguns and dashed down the mountainside, slipping, stumbling, falling. Going forward only when we heard the monkeys call, we were still a hundred yards away when a huge black gibbon leaped out of a treetop just as I stepped from behind a bush. He saw me instantly. For a full half minute he hung suspended by one long arm, his round head thrust forward staring intently; then launching himself into the air as though shot from a catapult, he caught a branch twenty feet away, swung to another, and literally flew through the treetops. Without a sound save the swish of branches and splash after splash in the leaves, the entire herd followed him down the hill. We returned to camp wiser in the ways of the gibbons of Ho-mu-shu.

Two days later we were sitting on a bed of fragrant pine needles, watching a squirrel, when the crazy laughing call of the gibbons sounded from across a deep ravine. A swirl of leaves showed us where they were and as one big fellow swung out on a branch with one arm, I knocked him off with a bullet from my rifle. A brown female ran up the tree trunk a few seconds later and peered down where the first monkey had fallen. I dropped her, too, and the rest of the herd faded into the jungle like flitting black shadows.

A clean wall seventy feet high fell away in a sheer drop to the mountain torrent which leaped and foamed over a chaos of jagged rocks, and separated me from the spot where the gibbons had fallen. I skirted the rock-face and had laboriously worked my way around and directly above it, when a vine stripped off and I began to slide. Faster and faster I went, dragging a mass of ferns and creepers with me, for everything I grasped gave way. Hardly ten feet from the edge of the precipice, the rifle slung on my back caught on a tiny ragged ledge and held me flattened against the cliff. How I managed to twist about and work backward, inch by inch, to safety among the trees, I only vaguely remember. At the bottom of the canyon, I felt weak and sick. It was half an hour before I could climb the mountain back to camp.

There were other hunts for goral, sambur, and leopard before we were back again on the Burma Road, working slowly southward to Bhamo and civilization. For nine months we had been wandering in the wilderness to bring out the biggest collection ever taken in Asia on a single expedition. There were 2,100 mammals and a thousand birds besides many fish, reptiles and batrachians, all from regions where no collector had ever before set foot. We came to Bhamo early in the afternoon and the resident commissioner offered us the hospitality of the circuit house. At the club that night, a military band was playing, and men in white, well-dressed women, and officers in uniform strolled about or sipped iced drinks beside the tennis court. We felt strange and shy, but doubtless we seemed stranger to them for we were newly come from a far country which they saw only as a mysterious unknown land.

Chapter 15

Wartime Interlude

In Bhamo, for the first time in many months, we had news of the outside world and heard that the United States had declared war on Germany. The moment I knew we were in the struggle for better or worse, I felt a tremendous urge to get home and into the show. But that was easier said than done. Down the Irrawaddy to Rangoon, across the Bay of Bengal to Calcutta and the whole breadth of India to Bombay we had to go to get ship for the United States. None would take passengers via the Suez and Mediterranean so we had to retrace our steps through the Indian Ocean to Singapore, China, Japan, and across the Pacific. A hot impatience robbed me of pleasure in the wondrous sights of India, and three weeks waiting, waiting, waiting in stifling heat at the Taj Mahal Hotel in Bombay drove me nearly wild. Cholera was discovered in the hotel and we got out just ahead of the quarantine officer. Every drop of water was boiled and every bit of food we ate was prepared by our own Chinese servant. If a fly buzzed in the room it was hunted down like a poisonous reptile.

At long last we reached New York with all our collections. And the day after Christmas my first son was born. He was named George Borup Andrews after his explorer uncle who had made a brilliant record with Admiral Peary on his expedition to the North Pole and whose tragic death was one of my greatest sorrows. Now George is a lieutenant in the U. S. Flying Corps. He got his silver wings the hard way, after spending a year as a volunteer in New York's illustrious Seventh Regiment.

Soon after George's birth I went to Washington to offer my services to the secretary of war, whom I had known for years.

"Strangely enough," he said, "I was just about to write you for I saw by the papers that you had returned. We need you here."

"But, Mr. Secretary, I don't want to fight the war in Washington. I came back to go to France."

"Yes, I suppose you did, but at this time each of us must do the job for which we are best fitted. We need you in the Intelligence Service. There are so few men who know the Far East as you do that we've got to have every one."

There wasn't much that I could say by way of argument but I left the secretary very much depressed. A desk job in Washington, even though I did wear a major's oak leaves, was an anticlimax to my dreams of active service. I walked over to the Cosmos Club for luncheon, feeling like a glass of stale beer, and sat disconsolately at a small table by myself. From across the room a tall man beckoned. It was the late Charles Sheldon, big game hunter and wealthy sportsman who was a friend of every naturalist.

"Roy, this is amazing. How did you get here so quickly? I only wired you this morning."

"I didn't get any wire. I came down last night and have just seen the secretary of war. Damn it all, he wants me in the Army Intelligence; I'll be here in Washington for the duration."

"But that's just what I wanted to see you about. I'm in the Naval Intelligence. We've got a job that's just made for you and it won't be in Washington. It will be in the Far East as fast as you can get there and I'll promise plenty of action. Come with me after luncheon to see the admiral."

A week later I was on my way back to China, bound specifically for Peking.

I arrived to find Peking a city of intrigue. No one believed that anyone else was what he said he was or that his presence in China did not have some underlying secret significance. New faces were continually appearing at the club or the Wagon-Lit Hotel, sending a flurry of gossip through the foreign community. Money was being made in undreamed of quantities in almost all neutral countries, and China's untouched wealth attracted the old rich as well as the newly rich as a candle does a moth.

Political chaos had reigned in China ever since the Revolution in 1911. The Republic was a joke. War lords, by courtesy called "military governors," controlled every province, each with his own army of conscripts, or brigands, and civil war went blithely on. When a war lord began to feel his oats he attacked his next door neighbor if he thought he could get away with it. No one suffered much except the peasants, for there was little killing. As Admiral Tsai Ting-kan remarked when someone commiserated with him on the civil wars in China: "You forget how very civil they are." During one abortive attempt to capture Peking, a newspaper reported: "There were ninety thousand rounds of ammunition fired today and one man killed. He was a peanut vendor."

Usually the inter-provincial affairs were settled by double-crossing or buying off the other fellow. Armies were bought and sold like merchandise. The central government in Peking was a farce yet it continued to function in a way which only an Oriental could understand. The graft and official corruption would have made a Tammany politician look like a suckling babe but it was duck soup to the foreign concession hunters who were legion. The whole atmosphere was one of excitement, gaiety and very little worry about the war, which everyone was certain would be won now that American troops were arriving in France by the tens of thousands.

After China's entry, one of the first acts was to corral most of the German and Austrian nationals and intern them in a hotel outside Peking not far from the Summer Palace. It was a useful precaution, I suppose, and it marked the beginning of the end of the foreigner's domination in the Far East. Peasants and coolies gathered outside the wire enclosure to gape at the Germans like animals in the zoo. For the first time in a generation, white men were completely at the mercy of the Chinese. Their superiority gained by guns, not culture, was gone. The old China hands drank their whiskey and sodas solemnly, prophesying dire results from the outrageous procedure. Even though the internees were Germans, they were still white men and no good would come of it!

Peking remained my headquarters all through the war period, but it was not a static existence for me. My peregrinations took me over much of China and Manchuria, twice across the Gobi Desert from Kalgan to Urga, the capital of Mongolia, and northward, on horseback, through the vast unexplored forests into Siberia. It was work I loved because most of it fitted into my peacetime job of Asiatic exploration. New country, new customs, new people every day.

Always time was of the utmost importance, so when I crossed the Gobi it was by motor car. Only two or three years earlier, the first automobile had been driven over the ancient caravan trail, fighting rocks and sand and mud but getting through. Instead of lurching back and forth between the humps of a camel for two months, one could do the journey in five to seven days in a car—if one were lucky!

The first time I crossed the desert was in the early autumn. It was an exciting trip and turned out to be a bit on the dangerous side. Mongolia wasn't officially at war with anyone so the only human danger we expected was from Chinese bandits in the cultivated area a hundred miles north of Kalgan. The first day took us out of that and we drove placidly along a hard trail just over the frontier of Outer Mongolia. Charlie Coltman was asleep in the back seat when we came to Ude where a great promontory juts out into the plain. Suddenly, five men appeared on the outcrop and opened fire. Why, God only knows, but the bullets were zinging above our heads and plunking into the open car every second. Charlie woke with a yell, frantically pulling our rifles out of the cases. Humped over the wheel, I zigzagged back and forth across the trail, but the lead still kept coming.

I leaned back to take my rifle and at that split second a bullet shattered the whole lower side of the steering wheel just where I had been leaning. It couldn't have been a narrower squeak.

The trail led around a high wall of rock into the sandy bed of a dry stream. The car stalled in three seconds, but at least we were out of sight of the men who were trying to kill us. Doubtless they knew we'd be stuck there; all autos were. Leaving the motor running, we climbed the rocks and peered over the top. The five men stood in plain sight. They were dressed in Mongol gowns but that didn't prove anything. They might have been Russians. I rather thought they were because their shooting was so good.

Soon our potential murderers started to climb down the cliff, evidently bent on finishing off what they had begun. But we weren't having any. Charlie picked one fellow silhouetted against the sky. I lined my sights on another in front. Bang, bang went our rifles. Charlie's client sat down suddenly and rolled over. Mine did a magnificent swan dive right off the cliff. The other three ducked back among the rocks. It must have been a bit of a surprise to them. I guess they thought we didn't have rifles since we had not returned their first fire. When we reached Turin and spent the night in a big Lama temple, I couldn't resist burning a couple of joss sticks in front of the god who grinned so benevolently upon us.

Three days later we got into Urga during a night black as the pit, and slept in the yurt of a Mongol Duke. When I went outside in the morning to douse my face in ice-cold water, I could hardly believe my eyes. I knew I wasn't dreaming, but by some miracle, time and space seemed to have been obliterated. I might have been back in a settlement of western America during the Indian fighting days. Every house was surrounded by high stockades of unpeeled logs and, in an open space by itself, just where it ought to be for proper defense, was the block house. (I learned later that it enclosed a Lama temple.) Larch trees straggled down almost to the village edge. They came from the unbroken reaches of the virgin forest stretching away in a vast rolling sea of green far beyond the Siberian border. The picture only needed a befeathered Indian or a buckskin-clad frontiersman emerging from the woods with a deer slung across his back.

I was keen to explore the rest of Urga, but my stomach rebelled at the thought of doing it in a motor car. Our Mongol host produced a pony and I rode for two miles along the road bordered by brilliantly painted Russian cottages, to where it debouched into a wide square and became a heterogeneous mixture of Mongol, Russian, and Chinese. Palisaded compounds, gay with fluttering prayer flags, ornate houses, felt-covered yurts, and Chinese shops, mingled in a dizzying chaos of conflicting architecture. Three races had met here and each maintained its own customs and ways of life. High above the city, dominant and overpowering, stood the great temple of the Living Buddha, surrounded by the pill-box dwellings of fifteen thousand Lamas. On the street were Mongols in half a dozen different tribal dresses. Tibetan pilgrims, Manchu Tartars, and camel drivers from Turkestan ate and drank and gambled with Chinese from civilized Peking.

I rode slowly about the square gazing at the barbaric splendor of the native dress. The women were fantastically arrayed in every color of the rainbow, with the hair done over a framework like the horns of a mountain sheep. Besides gowns and sashes of dazzling brilliance, the men wore high peaked hats of yellow and black, flat pancake fuzzy creations, and yellow helmets of hard, shiny lacquer. Suddenly, I heard a chorus of wild yells and down the main street dashed half a dozen Mongols, lashing their ponies, peacock plumes streaming, each with a rifle slung on his back. They were fresh from the plains for a day in town, even as the cowboys from our own great West. It seemed impossible that this could be the life of every day and not some special celebration, for it was like a pageant on the stage of the Hippodrome or a Wild West show.

Near the main square stood a double stockade of high unpeeled logs with pointed ends. It interested me especially, for screams were issuing from behind the gates. Someone told me it was the jail and the cries were the usual morning "prisoner's song" begging for food and water. Louis XIII of France, if I remember correctly, at one time was very pleased with himself because he had invented a new and particularly horrible method of torture. It was a cage so small that a man could neither lie down nor sit erect. He invited those gentlemen whom he particularly disliked to occupy these pleasant apartments. When they were safely inside, and he felt the need of relaxation, he used to visit the torture chambers to taunt his erstwhile friends.

Such, only a few degrees worse, was what I saw in the Urga jail. In the small rooms, almost dark, were piled wooden boxes, four feet long by two and a half feet high. These coffins were the prisoners' cells. Some of the poor wretches had heavy chains about their necks and both hands tightly manacled. Their food, when the jailer remembered to give them any, was pushed through a six-inch hole in the coffin's side. Some were imprisoned here for only a few days or weeks; others for many years or life. Often their limbs shriveled away and I saw one poor creature whose arms and legs were only skin drawn tightly over the bones. He had, he said, been there five years as nearly as he could remember. Even in winter, when the temperature drops to sixty below zero, they were given only a single sheepskin for covering. How it was possible to live in the indescribable filth, half fed, half frozen, and suffering the tortures of the damned was beyond my ken. The jail was a pretty gruesome picture but it was a reminder that Mongolia still lived in the Middle Ages even though the twentieth century was pressing hard upon its borders.

It was early autumn, the loveliest month of the year, when I first crossed the Gobi. The second time I came in the dead of winter in a temperature of fifty degrees below zero. We were in an open car, too, but dressed in sheepskins with face masks and goggles. The motor chugged away, never stopping during those bitter six hundred and fifty miles. We did not dare let it stop for it would have frozen instantly. It wasn't so bad at first, but for the last two days we faced a blizzard which swept the snow in wraith-like streamers across the open plains and piled white hills against the rocks. We stuck to the trail when possible, but drifts turned us off and that was dangerous. The ravines and gullies were level full. Sometimes, in an innocent-looking patch of snow, the car would drop from under us with a sickening lurch into the depths of nowhere. Then it was dig, dig, dig, until we were on firm ground again. Finally, we staggered into Urga so nearly frozen that it seemed the cold had reached the very marrow of our bones.

It is not permitted, even now, to tell what part I played in the war. Suffice it to say that it was exceedingly unimportant although interesting and at times exciting. But I was bitterly disappointed not to be in France. Everyone knew that the war must be won on the western front. Asia was just a side show to which only those who were there paid much attention. China had declared war on Germany chiefly at the instigation of the United States, through our minister Dr. Paul Reinsch. Her chief physical contribution could be labor battalions and many of these were sent.

But there were diplomatic reasons which made China's entry into the war very important. Japan didn't want China in for reasons of her own. Japan had viewed the picture pretty realistically before they went in on the side of the Allies even though the Anglo-Japanese Alliance required them to do so. It was a completely selfish decision. Germany had been her model not only for the military training of her officers and army but for the development of her internal affairs as well. All her sympathies lay with Germany. She was, however, convinced that the Allies would win in the end and that she stood a better chance of furthering her aims in China by going in on their side.

According to modern procedure, there was nothing unusual in this. Certainly no one can believe that in the last twenty years international honor has been more than an empty phrase. Treaties have become a diplomatic waste of time. When national expediency dictates, they are tossed into the fire. God forbid that it will always be so, and I have faith that the time will come again when international promises are considered as inviolate as those of an individual. But that time is not yet.

Japan wanted Shantung Province of which the German-held port of Kiaochow was the gateway. She took Kiaochow and then made passes at the Shantung railway, which resulted in severe fighting with the Chinese. But China's entry into the war threw a wrench into Japan's diplomatic machinery. She could hardly carry on open, armed aggression against one of her own allies! By bringing in Chinese bandits, which were legion, Japan endeavored to foment internal disturbances in Shantung so that she might say it was necessary for the good of the Allied cause to police the province with Japanese troops.

Along with the bandits, she tried another typical Japanese method, one so low and vile that probably Germany is the only other nation in the world which would resort to it. I mean the drug habit. Japan flooded Shantung Province with opium, cocaine, heroin, and morphine. The drugs were put out in cigarettes, in "Jintan: Long Life Pills," and distributed in a dozen other forms. She hoped so to weaken the character of the people that later her conquest would be easy.

Japan has done exactly the same thing in Manchuria since she moved in during September 1931. I have seen such terrible results of drugs over all the Far East that, to me, the national or individual purveyors of narcotics are the most despicable of all creatures. Murderers or kidnapers are gentlemen compared to them.

Even though Japan was an ally of the United States, all during the World War our diplomatic pouches were tampered with by them and it became necessary to send diplomatic mail across the Pacific only on American ships. I happened to be in Shantung Province when the Armistice was declared and it was laughable to see the consternation of the little yellow traitors. Japan's nefarious schemes against China had been nipped in the bud by the cessation of hostilities and the officials didn't know how to act until orders arrived from Tokyo. As a result, for many hours after the news reached us, all the Japanese remained indoors and the consulate was closed. Kiaochow was ablaze with flags, cheering crowds surged through the streets, and in every foreign house corks were popping. But not a sight nor a sound from the Japanese; just a total absence of our gallant ally, privately and officially.

Suddenly, the Japanese consulate and all their nationals burst into a forced frenzy of delight. Evidently orders had come from Tokyo "to rejoice and be exceeding glad." Japanese flags appeared like magic. Officials in top hats and Prince Albert coats, or gold-braided uniforms, rushed frantically to the consulates of the Allied Nations to bow and suck in their breath and with frozen smiles reiterate, "We are very happy—we are very happy."

When the Japanese consul-general said that to me, I could hardly refrain from calling him a liar to his face, for I happened to know rather intimately just what part he had been playing in obstructing the Allied cause. I knew, too, that in the Japanese consulate at that very moment hundreds of pounds of heroin, opium, cocaine, and morphine in cases marked "Ammunition" awaited distribution at the Japanese consul's command. Ammunition it was, of the most deadly and soul-wrecking kind.

And so I relieved my feelings considerably by grasping his little outstretched hand in a grip that made the bones crack. "I, too," said I, "am very, very, *very* happy." At each "very" I gave another squeeze. My hands are like iron and even Gene Tunney has never been able to hurt me in a grip. Smiling into the consul's eyes like a Judas, I saw them darken with pain and his face turn a pasty green. When I released his hand I thought he was going to faint but he gulped down a glass of brandy and walked unsteadily out of the house. The fact that I had crushed his hand could not have official repercussions because I was only a private American citizen of no importance who was polite and very, very, *very* happy that the war had ended.

A few days after the Armistice, I returned to Peking. The foreign residents were still *en fête*. A fantastic round of dinners, dances, and celebrations had stopped all ordinary occupations. It was a priceless excuse for uninterrupted pursuit of pleasure, and Peking residents were not the ones to miss such an opportunity when it came knocking at their doors.

After one pretty wet party at the club someone suggested that, in fairness to ourselves, we ought to pull down the von Kettler monument. It seemed a sound idea at the moment and it was not the time to remember that the von Kettler monument, on Hatamen Street, was a massive arch of solid marble. To destroy such a structure would require charges of dynamite, not mere ropes and human strength. But champagne was buzzing in our brains. As I remember half a dozen lines, hardly more than string, were attached, we surged valiantly, the strings broke, and all of us were precipitated on our bum-fiddles in the gutter, much to the delight of hundreds of Chinese spectators. Feeling very silly, back to the club we went to drown our discomfiture in more champagne and tell each other how it should have been done. A few weeks later the "Chinese Government" with official sanction of the august Diplomatic Body, took down the arch, bit by bit, and reassembled it in the public garden of the Forbidden City as a "Victory Memorial."

Von Kettler, by the way, was the German minister who, when the Boxer Rebellion of 1900 was about to break, persisted in going unaccompanied to the foreign office even though he had been warned that it was virtually suicide. He was killed and after the Boxer trouble the German government demanded that the Chinese erect a memorial arch, or *p'ailou,* on the spot where his murder took place, as a perpetual reminder of national humiliation.

Chapter 16

Dog Eats Man

Those two glimpses of Mongolia which I had during the war made me know that I would surely return to this "Land of Yesterday." And so it happened, just as I planned, for after the Armistice, in the spring of 1919, I came back again. I had to use a motor car to Urga for the sake of time; then with camels, ponies, and carts I went to the desert as the Mongols go.

I knew from the first that Mongolia was where I would make the scientific attack on Central Asia to test Osborn's theory. My look at Tibet from its edge showed it wasn't good enough. The plateau was too high, the political difficulties too great, and the religious superstitions well nigh insurmountable barriers. But Mongolia had all the answers and I needed only to know it intimately. As usual, Professor Osborn had agreed to furnish half the funds if I would raise the rest. I did my part by letter. Loyal friends pledged enough money to make possible five months in the Gobi. It was to be a zoological expedition, of course, bringing to the Museum new collections of mammals and birds from an almost unknown country.

Never, I think, have I been happier in the field than during that summer in Mongolia. There were no details or people to worry about; I could go where I pleased, enjoying every campfire and every day of brilliant sun. And I came to like the Mongols. They were wild, independent folk, hard living, virile, meeting life in the raw and asking no quarter; as untamed as the eagles that soared above their yurts. The cowboys of our own early western days were their counterparts. Similar conditions had bred similar races of men. In strength and endurance I believe the Mongol of today is just as good a man as were the warriors of Genghis Khan. As a matter of fact, their peacetime life hasn't changed much, for the country hasn't changed. They were always nomads and will remain so until the last of their kind is gone.

It took a bit of doing to become accustomed to their smell. One Mongol in a closed room is equivalent to one skunk in the same space, for they never take a bath from birth to death, except by accident. There isn't much water and a Mongol doesn't see the need of cleanliness any more than he sees the need of chastity in his women.

They are unmoral rather than immoral. A man may have only one wife, but keep as many concubines as his means allows, all of whom live in a single room of the yurt. A girl remains a virgin only by inclination and that is seldom long. Hospitality is a law of the land extending even unto the offer to a visitor of a man's wife or daughter as a companion for the night. No one's feelings are hurt if the offer is refused. It is just like saying you don't want a cup of tea. Of course, this promiscuity leads to disease and I am sure it is not too much to say that ninety per cent of all Mongols are afflicted with venereal trouble of some kind. They seem, however, to have developed a degree of immunity to syphilis, and a missionary doctor in Inner Mongolia told me that he often saw normally healthy babies whose parents were both syphilitic.

When I first went there, Outer Mongolia was governed by the Hutukhtu, the Living Buddha, third in rank in the Lama hierarchy. First, of course, was the Dalai Lama, second the Tashi Lama, both of Tibet; then the Hutukhtu of Mongolia. The Hutukhtu was an extraordinary individual who took his exalted position seriously enough, I am sure, but wasn't adverse to having a spot of fun on the side. In his younger days, he was a "bit of a bird" loving inordinately wine, women, and song. The stories of midnight revels in the Hutukhtu's Palace were still rife when I first went to Urga.

Somewhere he had obtained a Sears Roebuck catalogue and
his days were spent poring over its offerings with a translator at
his elbow. As a result, his palace, beside the Tola River, just under
the shadow of "God's Mountain," was a veritable junk shop of
Western inventions which had amused him for a moment, only to
be tossed aside when he had become tired of them.

Like all Mongols, he had a keen sense of humor. Way back
in the early 1900's a motor car had been brought by camel to
Urga. I saw it in the "garage." It was a strange French thing like
an overgrown baby carriage, with a carburetor as large as a ten-
pound bomb which projected from the left side on a steel frame.
The Hutukhtu never rode in it but he did enjoy using its battery for
"blessings." Two wires were strung from the car, his clients would
grasp them reverently, and when all was set the Hutukhtu turned
on the current. No one doubted that he had been blessed!

I saw the Hutukhtu only once and then he was old and almost
blind. In return for my present he gave me a beautiful white jade
polar bear with ruby eyes which the prime minister told me had
been presented to His Holiness by the czar of Russia. Today it
stands on the piano in our house at Pondwood Farm.

I stayed only half the time on the plains during that first summer
in Mongolia and then went up into the larch forests hunting
moose, bear, wapiti, roebuck, lynx, musk deer, and wild boar.
Those northern forests and meadows in late July were something
to dream about. Never have I seen such wild flowers! Bluebells,
their stalks bending under the weight of blossoms, yellow roses,
forget-me-nots, and a dozen other flowers made a bespangled
carpet under foot. But I liked the wild poppies best of all for their
delicate fragile beauty was wonderfully appealing. I learned to
love them first in Alaska where their pale, yellow faces look up
happily from the storm-swept hills of the Pribilof Islands in the
Bering Sea.

The summer was over all too quickly, and I started back to China
in a motor car, but almost didn't get there. Our first night's camp
was at Turin, the core of a great mountain rising in a chaotic mass

of ragged spines and peaks above the plain. Only a mile away a lama monastery nestles in the heart of a shallow basin. There are, perhaps, five thousand priests living in tiny wooden shacks about the temples. Also dogs—hundreds of them. Great black shaggy fellows like the Tibetan mastiff, which slink about the priestly village waiting for scraps and the bodies of dead lamas.

It is a custom of the Mongols not to bury their dead. As soon as life has departed, evil spirits are supposed to take possession of the body, and it isn't a thing to have about the house, or even touch. Usually it is placed gingerly on a cart, and the driver dashes off at full speed, hoping to get rid of his unwelcome burden at some place during the journey. Where, he doesn't want to know, for it is bad luck to look back. The dogs, wolves, and ravens do the rest. I remember timing a pack of dogs at work on a dead lama. It took them just seven minutes to tear the body into a hundred bits.

As a result of this diet of human flesh, every person is a potential meal to the Mongol dogs. Very dangerous they are and no one ventures near a yurt or walks on the streets of Urga at night without a heavy club or weapon of some sort.

When we camped at Turin, I didn't think about the dogs at the temple a mile away. As usual, we simply spread our fur bags side by side on the ground and went to sleep. Two loaded rifles were beside me, one a tiny .22 caliber and the other my 6.5 mm. Mannlicher. During the night my companion was restless and at two o'clock sat up suddenly, wide awake. There, circling about us in the moonlight, were a pack of fourteen dogs. A scream brought me up just as they dashed in. Half awake, I grabbed the first rifle my hand touched and fired blindly. It happened to be the .22 caliber Winchester but the tiny bullet must have caught the big leader in the head for he sank in his tracks, stone dead. The pack swerved, swept by only a few yards away, and I fired twice more, hitting two other dogs. Instantly there was a bloodcurdling chorus of yelps and growls as the wounded animals were devoured alive. I dragged the dead leader far beyond our camp and the next morning all that remained of his carcass were a few bits of bloody hair. Had my companion not waked at that very second we never would have lived to see another day.

Back in Peking, I took account of stock. The summer's work had brought a collection of fifteen hundred mammals, all from a region that was virtually new to science. That was important as a tangible result to the Museum authorities and my friends who had helped finance the trip. But the really vital thing was the knowledge I had gained of Mongolia as a theater of work for my great expedition. Every detail was clear in my mind and I could hardly wait to get back to New York.

The first ship sailing was the old *Empress of Japan,* a yacht-built vessel now long out of service. Her run was the northern route and we got everything the elements could hand out on that trip home. Fog, rain, sleet, snow, and gale after gale climaxing in a hurricane that lasted three days. Most of the time water was sloshing in the cabins. I've never liked the sea, although much of my life has been spent on ships, and I got more than enough of it on that voyage.

I was saddened at Vancouver by news of the death of my old friend Admiral Peary. He had done a lot for me in encouragement and inspiration and I had looked forward to talking over with him in Washington the plans for my big expedition. The world never knew how great a man he was!

It was three days after reaching New York that I had luncheon with Professor Osborn in the president's office at the Museum. He knew that I had something important on my mind but it was characteristic of him to keep business until the meal was ended.

"Your stomach and your head can't both work well at the same time," he often said. "Eat first and then think."

When coffee had been served and we were smoking comfortably, he smiled: "Now let's have it, Roy. It's another expedition, I suppose."

"Yes, that's why I came back. The expedition I've been dreaming about for years. To test your theory of Central Asia. Especially, to try to find evidence of primitive man. Mongolia is the place."

"Well," said the President, "how do you propose to go about it?" That was my cue. I began talking as I never had talked before. In two minutes, Professor Osborn's eyes were glowing. He stopped smoking and just sat there looking hard at me and absorbing every word.

"More than that, we should try to reconstruct the whole past history of the Central Asian plateau—its geology, fossils, past climate, and vegetation. We've got to collect its living mammals, birds, fish, reptiles, insects, and plants and map the unexplored parts of the Gobi. It must be a thorough job, the biggest land expedition ever to leave the United States."

"Of course," said he, "it's a gamble so far as fossils are concerned. The Russians never found anything in Mongolia. What makes you think you can do better?"

"Their past work has been too much political and too little scientific. There have been a few good men—Prjevalski, Kozlov, Obrechev—but along with the others they had to produce economic or political results. Science never was the primary aim of any of the Russian expeditions. No one has attempted to do it the way I plan. Moreover, they all used camels. They could only average ten or fifteen miles a day. We'll use motor cars. We can go a hundred miles a day, if my estimate is right, so we ought to do in one season quite as much as the others did in ten."

"How do you know you can use motor cars in the Central Gobi?"

"I don't *know* it. But I believe from all I have seen of the country that it can be done. It will be largely a matter of preparation. The terrain is mostly fine gravel. I don't think there is much loose sand. We must have every conceivable motor part and experts who can almost build a new car if necessary. Such men exist. I can get them."

"How are you going to supply your cars with gasoline? You can't carry enough to go very far."

"We will have a supporting caravan of camels. It will act exactly like the supply ship to a fleet at sea. The camels must leave months ahead of us during the winter, and we'll meet them at a rendezvous perhaps six or seven hundred miles out in the desert. They will carry food, too, and most of our equipment."

"What about your scientific staff?" asked the Professor. "How do you intend to work?"

"We must have the best men in the world representing all the sciences that will help us solve our problem. Each man will assist the others with his special knowledge in interpreting what we find."

"But," said Professor Osborn, "you probably can't all work together in the same place at the same time."

"I've thought of that. There will be three or four separate units, each complete and able to maintain itself for at least a fortnight. The main camp will be the base and each party will work out from there."

The President asked me a lot more questions but I had all the answers ready. Why not? I'd been thinking about it for months in the Gobi and out of it. Finally he said: "Roy, we've got to do it. The plan is scientifically sound. Moreover, it grips the imagination. Finances are the only obstacle. You estimate five years for the expedition and a quarter of a million dollars. That's a lot of money and there is a severe business depression at present. Of course, the Museum will do all it can but that won't be much in the way of cash. Getting the money will be up to you. What are your ideas on that score?"

"My only chance, I believe, is to make it a 'society expedition' with a big S. You know that New York society follows a leader blindly. If they have the example of someone like Mr. Morgan, for instance, they'll think it is a 'must' for the current season. 'Have you contributed to the Roy Chapman Andrews expedition? If not, you're not in society.' That's the idea."

"Yes," said the Professor thoughtfully. "It might work. Anyway, it's worth trying and I think the best way. This is big enough to be interesting to those people who are accustomed to think only of big things."

"Then if it's all right with you, I'll get to work at once—tomorrow."

The Professor's face broke into the wonderful smile of affection and enthusiasm which made his friendship one of the most valuable and inspiring things that ever came into my life.

"In spite of all the difficulties, Roy, I'm sure you can do it. Whom are you going to see first?"

"Mr. Morgan," said I. "If he is interested, I'm sure of the others."

Chapter 17

Wall Street Ramblings

The next day I telephoned Mr. J. P. Morgan at the bank and asked for an interview. The day after that I met him in the Morgan Library, unfolded a map and we bent over it together.

There is always something exciting about a map and this was particularly true in those days when a lot of blank areas were still marked "unexplored." The entire Central Gobi was a white space with only a few thin lines waving uncertainly across it. I launched into my story with the enthusiasm of a fanatic. In two minutes, everything was forgotten except the prospect of what could be done if only I had the money. Mr. Morgan listened with rapt attention. Two or three times, he interrupted with questions: "How are you going to get there? Why do you think there is anything there?"

I told him how the camel caravan carrying food and gasoline would be sent months in advance during the winter to a rendezvous a thousand miles out in the desert. I told him how my staff in the motor cars would follow early in the spring, and after we had made contact with the camels I would use them as a movable supply base, like the mother ship to a fleet of submarines. I explained Professor Osborn's theory of the distribution of animal, and particularly human, life from a Central Asian point of origin.

"It is only a theory, you understand, but it seems to me that the reasoning is sound. Anyway, we'll never know until we try to find what's there. Someone else will do it if we don't."

At the end of fifteen minutes, I stopped, breathless. Mr. Morgan swung about with his eyes aglow.

"It's a great plan, a great plan. I'll gamble with you. How much money do you need?"

"A quarter of a million dollars and five years at the least, Mr. Morgan."

"All right, I'll give you fifty thousand. Now you go out and get the rest of it."

"That's wonderful. I'll get it. There is, I suppose, no use asking who of your friends might be interested?"

"No, I couldn't tell you that. But hold on. Albert Wiggin! He sent a man to me last week. It cost me ten thousand. You go to him, tell him I sent you and that he'll do damn well to shell out."

I didn't know Mr. Wiggin, who was president of the Chase National Bank, but I called his secretary, asked for an interview, and said I had a message for him from Mr. Morgan.

A few days later I was ushered into his office. Mr. Wiggin sat behind a glass-topped desk. We chatted for a moment and then the banker said: "My secretary told me you had a message from Mr. Morgan. May I ask what it is?"

I smiled. "Shall I give it to you in his exact words?"

"Yes, of course."

"I'm trying to finance a great expedition to Central Asia and Mr. Morgan has given me fifty thousand dollars. He told me to come to you, say he sent me, and that you'd do 'damned well to shell out.'"

Mr. Wiggin slapped his desk and laughed, albeit ruefully. "That would be the chap I sent to him last week. Well, tell me about the expedition."

I got out my map and started but it was obvious Mr. Wiggin wasn't much interested. So I cut it short and rose to go.

"Very exciting, very exciting," he said. "I'll send you a check in a few days." He did, but it wasn't for ten thousand!

The late Arthur Curtis James, holder of more railroad securities than any other man in the world, was next on my list. Mr. James was one of our trustees and I had met him frequently at the Museum. His secretary gave me an appointment but she picked a bad time.

At nine o'clock in the morning I went into his office in the Phelps Dodge Corporation in Wall Street. Mr. James was hopping mad. He had just read a letter from Professor Osborn asking for a contribution to the Museum's deficit of seventy-five thousand dollars. As I came in the door, Mr. James, without any preliminary greeting, roared: "Here's a letter from your president. He says he knows I'd enjoy contributing to the deficit. I can't think of anything I'd 'enjoy' less! There shouldn't be a deficit. He ought to spend his income and no more. I'm going to get off the board if he uses his trustees just as check books."

There didn't seem to be much that I could say since the Museum finances were not my responsibility.

"I'm sorry, Mr. James. I seem to have caught you at the wrong time. I was going to talk about an expedition to Central Asia, but I guess it had better wait."

"No, no you're here. Go ahead. What's it all about?"

I began rather lamely, I'm afraid, for it was a tough spot. I couldn't capture his interest and his eyes were still snapping with anger. Finally he broke in: "How much am I going to be stuck for this expedition?"

That made *me* mad.

"Mr. James, you're not going to be stuck a damned cent. This isn't something the trustees are doing alone. I came to you as an individual. If you aren't interested, I don't want your money. You are busy, and so am I. Good morning."

I got to my feet, reached for my hat and stick, and started out the door. The anger went out of his eyes as though a switch had been turned. He had never called me by my first name before, but he said: "Now, now, Roy. I'm sorry. I shouldn't have ripped out at you like that. I can't let you go off to China being sore at me. Will ten thousand do you any good? You can have it and welcome.

Teach me not to lose my temper. Always ought to pay for it when I get mad."

I walked out of the Phelps Dodge offices in a daze. Raising money from the world's greatest financiers surely was a strange business. One man gave it to me because it was the pay-off for another and now I had ten thousand by getting mad and making a friend ashamed of his own temper. Talk about adventures in the Gobi Desert! Adventures in Wall Street were just as exciting.

About this time, I met Louis Froelich, editor of *Asia* magazine, which was a baby of the late Willard Straight's. Mr. Straight, whom I had known in China, had married Dorothy Whitney, daughter of the great financier W. C. Whitney. Louis had the idea of a hook-up between the magazine and my expedition and it made sense. He would try to persuade the American Asiatic Association, of which *Asia* was the official organ, to donate some money to the expedition and in return I'd give the magazine first call on anything I wrote. Louis took me up for cocktails with Dorothy Straight, whom I had never met. She charmed me at once, as she did everyone else, and entered enthusiastically into the plan. She would donate five thousand dollars personally, she said, and, what was more important, give a dinner and reception to start off the expedition in the social world. With J. P. Morgan, Albert Wiggin, and Arthur Curtis James already pledged as backers, I was riding the crest of the wave.

Dorothy Straight's house on the night of her dinner was a blaze of lights and flowers. Twenty people were at the table among whom I remember Thomas Lamont and Dwight Morrow, of J. P. Morgan and Company, William Boyce Thompson, the international mine owner, Clarence Mackay, president of the Postal Telegraph Company, and the distinguished snowy-haired Mrs. E. H. Harriman. While we were having coffee, thirty or more other guests assembled in the ballroom.

With the aid of lantern slides from my last trip to Mongolia, I tried to give them an idea of the Gobi, of how I planned the expedition, and what were the chances of success. Always I stressed the point that it was a gamble. No one knew what was there, if anything, but the scientific dividends would be enormous if there were any dividend at all. I talked for an hour, then an orchestra began to play and there was dancing until morning.

The evening was a great success so far as the expedition was concerned. I hardly danced at all for little groups of people kept me talking and I had the opportunity to tell my story individually to a dozen men whom it would have been difficult to interest in their offices during a business day. Of course, I didn't ask anyone for money.

"How are you going to finance it?" was an invariable question at some time in the conversation. I just looked shy and said I hoped my friends would be enthusiastic enough about the idea to help out with dollars. No one rushed up right then and there and said, "Let me give you a thousand."

But the immediate result of the evening was a flood of invitations to dinner at dozens of New York's greatest houses. Almost every night I donned white tie and tails, climbed into a taxi, and told my story at some function where the cards read: "To meet Roy Chapman Andrews." Rushing about all day, staying up until one or two o'clock every morning, keyed to a nervous pitch, almost wore me out at times, but never quite. Sometimes it seemed that I'd give my soul to eat crackers and milk and go to bed for just one evening instead of dining on caviar, green turtle soup, and roast duck.

The results, however, were worthwhile. My original idea that the expedition could be financed, even though there was a depression, by making it the society "thing to do" for that winter proved to be right. I asked very few people for money myself, but I did get others to do it for me and the checks came in.

There were no very large contributions, except Mr. Morgan's and Mr. James's. A number were for a thousand dollars. That is quite a lot of money to give but it is only a small dent in a quarter of a million. The fund was still a long way from the necessary figure and time was passing.

Professor Osborn and I decided to give a bang-up men's dinner at the University Club. We went over the list carefully and there were fourteen shining lights of American finance gathered in the beautiful Council Room where more expeditions have been launched, I think, than in any other place in New York City. The invitations went out in Professor Osborn's name and I was guest of honor. At my right was J. P. Morgan and on the left Judge Gary, president of the U.S. Steel Corporation. A little down the table sat George F. Baker, Sr., John T. Pratt of Standard Oil was there with Sidney M. Colgate, Childs Frick, Cleveland H. Dodge, and H. P. Davison. Probably there was enough money represented in that room to buy a quarter of the United States.

One thing that impressed me was what a good time all the men seemed to have and how completely carefree and charming they were. To see these titans of business at play was a revelation and a privilege. I had begun to realize that the average financier is an adventurer at heart although he probably doesn't know it. Making his money has been an adventure for him, and if you can give him an adventure in spending it, he'll have a lot of fun. But that was in 1920, remember, before war and taxes took their toll of the great American fortunes.

While we were having coffee, Professor Osborn introduced me and I gave a talk on the expedition plans with my Mongolian lantern slides. It seemed to go well and the men fired questions at me like machine-gun bullets for nearly an hour afterward. As he left about eleven o'clock, Mr. Baker said: "Come to see me at the bank any day. I'd like to talk with you some more. Bring your map. I love maps."

A few days later I sat beside Mr. Baker in his office in the First National Bank, with my map spread out. Mr. Baker pored over it like a schoolboy, asking me question after question about the transport and every detail of the plans. Several times I started to leave, but he said, "No, no, please don't go unless you have to."

At the end, he called a secretary and said, "Draw a check in Mr. Andrews's favor for twenty-five thousand dollars."

It was a magnificent help, but we were still short. The man I wanted to see next was John D. Rockefeller, Jr., but Professor Osborn, who knew him well, was discouraging.

"I've tried for years to get him to come to the Museum," he said, "but he never would. I don't think you've got a chance even to see him."

I made up my mind to try anyway, but not by the direct approach. A distinguished lawyer, J. Starr Murphy, was an intimate friend and confidential adviser of Mr. Rockefeller's and for that reason I had suggested that Mr. Murphy be included in our dinner guests at the University Club. During my talk, I watched him closely and saw that he was keenly interested. Two days later, I asked for an appointment at his office.

He had assigned me fifteen minutes, but I was with him more than an hour. Mr. Murphy was a scholarly man, deeply interested in evolution and the possibility that we might find early human types in the Gobi Desert. Finally, he said: "I'll talk to Mr. Rockefeller. Of course, I don't know whether or not he will be interested, but it is possible he might be very much so. Anyway I'll let you know."

About ten days later Mr. Murphy telephoned that Mr. Rockefeller would like to see me. I never had met, or even seen, John D. Rockefeller, Jr., and was enormously impressed by the simplicity and kindly graciousness of his manner. I talked for half an hour, but he said very little. In contrast to Mr. Morgan, who was obviously thrilled by my story, Mr. Rockefeller showed quiet interest but it might have been merely the courtesy he would extend to any guest. I left without the slightest idea of whether or not I had made a favorable impression. Imagine my excitement, therefore, when, a fortnight later, I received a note from his secretary: "Mr. John D. Rockefeller, Jr., directs me to inform you that he will be pleased to contribute fifty thousand dollars to the Asiatic expedition. He wishes to know how the money shall be paid. This is a personal contribution and will not go through the Foundation."

Professor Osborn was even more excited than I because he thought it might mean an awakening interest in the American Museum and a gift from the Rockefeller Foundation. As a matter of fact, that is exactly what did happen a few years later.

With Mr. Rockefeller's gift to the expedition, the funds had reached the two-hundred-thousand-dollar mark, and I felt it was time to make a public announcement since we could be fairly certain of raising the remaining fifty thousand. Two months earlier I had written an article for *Asia* magazine detailing the plans of the expedition, and we carefully synchronized the press release with the day the magazine appeared on the newsstands. The afternoon before, I had a press conference at the Museum at which twenty-one reporters were present. We hoped the announcement would make the front pages in the morning and sure enough it got the place of honor in every New York newspaper, and the press wire agencies sent it over all the world. That pleased us enormously, for it isn't easy to land on the front page of a New York daily. It meant the editors realized it was an expedition of considerable importance, and public interest was assured.

I was, however, greatly disturbed because every paper stressed the search for primitive man and said comparatively little about the broad scientific aspects of our plans. I was afraid the "missing link" angle would turn conservative scientists against us. In vain did I try to direct attention to the larger aspects of the work: the test of Professor Osborn's theory of Central Asia as a theater of mammalian evolution.

We told everyone that we did not know whether we would find human remains; that we could only hope. We told them that it was much like looking for the proverbial needle in the haystack; that human bones are so fragile they are not as readily preserved as are those of larger mammals; that even in the early stages of his evolution man was more intelligent than the animals around him and that he was not as readily trapped in bogs, quicksand, and rivers where his remains could be fossilized. We made it clear that the best scientific opinion pointed to Central Asia as the place of human origin, without any proof to support the theory, and that all we could do was to concentrate upon the problem in a thoroughly scientific way; that we should have to reconstruct the past climatic and physical conditions of the great plateau before we could feel we were working in the right place.

All this didn't create a ripple in the newspaper world. Primitive man was what they wanted and anything else bored them exceedingly. In a week, we were known as the "Missing Link Expedition."

When a phrase catches the public's imagination, it goes like fire in dry leaves. It was beneficial, in a way, for the man-in-the-street took a much greater interest in our expedition than he otherwise would have done. I was flooded with letters like: "Why go to Asia to hunt the Missing Link? I saw him in the subway this morning." That was the standard joke but there were dozens of others. One woman telegraphed: "Regarding search for Missing Link, Ouija Board offers assistance."

None of us had dreamed that the publicity would be so overwhelming. We were far from pleased by the line it took. We wanted to do a straight job with no ballyhoo but apparently it wasn't going to be possible. The number of prominent names on our list of backers coupled with the fact that the expedition was being undertaken under the auspices of the greatest scientific institution of its kind in the world, made wide publicity inevitable. The direction it took was a natural result of newspapermen's romanticism.

I had selected my scientific staff long before the public announcement, but we had a veritable avalanche of applications to join the expedition. Before it ended, these numbered ten thousand, including about three thousand from women. Since all except the scientific staff would be native Mongols or Chinese, I could only say no to the requests. At first I tried to answer the letters personally, but soon that became impossible. I had to resort to a printed card saying that the staff had already been selected but their application would be put on file for future reference.

Some of the letters were amusing. One day I heard my secretary exclaim under her breath, when examining the mail, "Why the very *idea!*" Then she remarked, "I don't know whether you will consider this amusing or not, but you had better read it and here is the photograph." It read:

Dear Doctor Andrews:

I want to apply for a position of secretary on your next expedition. I have written two books, but they have not been accepted yet.

I am looking for something occult and stirring and I think I can find it with you. I have seen your picture in the newspapers and I know from the kindness and nobility of your face that you know how to treat a lady.

If no position of secretary is open, perhaps you can take me just as a woman friend. I could create the home atmosphere for you in those drear wastes. I am sending my photograph but it is much better for you to see the original. How would Friday afternoon do for tea? After that I will leave it in your hands to judge.

Another woman wrote:

I am a lady who wants to join your expedition in any capacity. I am thirty years old. If you should answer please don't try to tell me of the dangers due to sexual feelings because they do not exist in me and cannot be aroused so that ends that argument.

Many of the letters came from ex-army men, particularly fliers. Most of them began: "I can't settle down to office work after the war. I want to get away where there is some excitement."

Teen-age boys by the thousands wrote. One implored me as follows:

Dear Mr. Andrews:

Mr. Andrews, I want to ask, nay plead, to accompany you on your dangerous trip. I say dangerous because I realize the risk of being killed by wild cannaballs and animals.

I can climb trees and am not light or dizzy headed and I am in good physical condition and would not cause you any trouble on that account.

I would make a good scout and I could work and help you in many ways and would gladly take the risk with you and if I was killed why of course that would be my hard luck.

I have all my life been interested in ancient animals and birds and the old time clothes that people of long ago wore. Please, oh, please Mr. Andrews allow me to go with you. I shall almost hold my breath till I get a reply from you and if you possibly or impossibly let me go with you say yes.

Sincerely,

P.S. Who knows, I might even save your life—of course probably I wouldn't but if I got the chance I would.

A man, obviously a German, wrote:

I have known that, organized by your great institution, soon from New York is to depart a hunting expedition to the Gobi Desert. If you need a barber and hairdresser (also sharp-shooter) I would be very glad to obtain that position.

Another was certain that he would be invaluable because he was a butcher by trade and could act as my personal bodyguard. Also a waiter said he was bored with waiting on table in a restaurant and would like to transfer his activities to our table in the Gobi. As an extra inducement he added, "I own a tuxedo. So you will not have to furnish that."

A lady from St. Louis wrote that after reading the news of the expedition, and its search for the "missing link," she had communicated with certain spirits, with whom, I judged, she was on familiar terms. They had told her of a buried city in the Gobi where a record of man's development might be found from the time he crawled on all fours to the dawn of history. The letter was so amusing that I asked her if she would inquire from her friends, the spirits, about the latitude and longitude of the city as the Gobi is rather a large place. A fortnight later she replied that the spirits were annoyed at my request but had vouchsafed the information that the spot was marked by four large stones half buried in the sand.

There were, of course, the usual "crackpot" letters. It is amazing how many people are loose in public although obviously insane. Probably their friends call them "queer," but crazy as ticks they must be to have written some of the epistles I received. The religious issue of human evolution got its share of attention, mostly from small-town preachers, but much less so than I had expected. My mail furnished an illuminating cross section of the kind of people who inhabit our great country and on the whole it was pretty good.

The last fifty thousand dollars to complete our quarter of a million dribbled in much too slowly for my sailing date and then the Lord tempted me. The vice-president of a great oil company said if I would take one of their geologists, who would act entirely under my orders but still do a spot of looking for oil, they would give me the fifty thousand. A similar proposition came the next week from a mining syndicate interested in the possible mineral resources of the Gobi. The offers were tempting because both concerns were of the highest reputations but I turned them down at the first interviews in spite of the fact that the hundred thousand dollars would have put our finances far over the top.

I was determined that the expedition should be strictly scientific and its objectives be only what we said they were without taint of commercialism. There could be no "secret covenants secretly arrived at." I thanked my Lucky Star a good many times later that we had nothing to conceal; otherwise the expedition would not have reached the Gobi. It never occurred to me that because we had for backers such men as John D. Rockefeller, Jr., J. P. Morgan, Cleveland H. Dodge, and other names synonymous with vast oil and mineral interests that the expedition would immediately be under suspicion, not only abroad but at home. All the contributors had given their dollars purely in the interests of science and exploration with no strings of any kind attached. Yet both the Chinese and Mongol governments refused to believe it, at first, in spite of the fact that it was an American Museum expedition with the blessing of the State Department and the president of the United States. I realized how difficult it is for such men to do anything, even though it is completely public-spirited, without being suspected of ulterior motives.

By the time we were ready to sail for China in March 1921, the money for the expedition had all been raised and the equipment shipped. Also I was well nigh a nervous wreck. It took several months of oriental quiet to put me right again, but after all, it was worth the price. The expedition of my dreams was an accomplished fact.

Chapter 18

My Peking Palace

Not since 1900 had there been such a dust storm as that which ushered us into Peking on April 14, 1921. The yellow blanket reached as far south as Shanghai and hovered over the sea sixty-five miles beyond the coast. It came from a land parched by fourteen well-nigh rainless months which had taken a heavy toll of human life. We could hardly see the great Tartar Walls as the train came into the station and for days afterward the air was like a London fog. The Chinese are very superstitious and we were told that no good could come in a summer which began with such a dusty spring. It was a bad omen—it meant famine, war, disease, and death.

The foreign community was always more or less affected by the Chinese superstitions and we were greeted with a flood of rumors. Peking was certain to be attacked and looted—even the day and hour had been set; no one could go into the interior, smallpox was raging, it would be dangerous to do this and dangerous to do that!

The same dear old hysterical Peking! We were rather a small community and excitement was a *sine qua non*. If no political bomb was ready for explosion, something must be manufactured to furnish conversation at the club and on the roof garden of the new hotel. So with dust, war, and smallpox we felt the summer was beginning rather well.

It was impossible, of course, to consider an expedition to Mongolia for the first year. There was too much preliminary work to be done. First, I had to find a suitable house as headquarters of the expedition; then arrange the complicated diplomatic negotiations for our work both with the Chinese and Mongol governments; buy camels, hire a native staff, get food and equipment packed, and the caravan started during the coming winter so that it might meet our motor party a thousand miles out in the desert.

The infinite details kept me busy as a bird dog but I found time to spend three hours a day at the North China Language School studying Chinese. Only two members of the staff had come out with me. Walter Granger, chief palaeontologist and second-in-command, was to spend the winter on the Yangtze River hunting fossils; the other scientists would arrive the following spring just in time to leave for Mongolia.

I was fortunate in finding an ideal house. Its former tenant, my old friend Dr. George E. Morrison, was one of the best known Britons in North China. His magnificent library, his brilliant writing for the London *Times,* his fascinating personality, and his interest in science and exploration made his home a mecca for travelers of every nationality. The palace belonged to a Manchu prince, all of whose other possessions had been taken from him when the emperor was dethroned in 1911.

A Chinese "house" is a series of separate one-story buildings connected by covered galleries and built around open courtyards. Our compound enclosed more than an acre within its walls. Space wasn't any object in the old days and two- or three-story buildings just wouldn't work in a Chinese ménage because of the concubines. Our particular prince had rated thirty-one concubines. He didn't really need that many for he took the last two or three at seventy years of age, but "face" demanded it. When he visited the emperor on formal occasions, all the concubines went too. They remained in one of the outer courtyards, drinking tea and cracking watermelon seeds between their pretty teeth, but they swelled the retinue, which was most important.

The Chinese ideograph for "trouble" is two women under a roof. Therefore, one can imagine what thirty-one would mean for any man. The problem was solved in the simple Chinese way by having a dozen or more courtyards and scores of rooms. If four or five concubines got on well together they shared the same courtyard off which opened their own little cubicles. Some of the ladies, who guaranteed fireworks when they met others of the family, might live for years without going into parts of the house where their particular "hates" resided. It worked out beautifully, for the prince's mother was boss of the household and ruled the little butterflies with an iron hand.

When Dr. Morrison took over our house there were ten courtyards and one hundred and sixty-one rooms. He knocked out partitions and pulled down two buildings, reducing the space to forty-seven rooms around eight courts. Chinese houses have stone floors and, of course, no modern plumbing. Morrison had only started on the reconstruction job when he died so I inherited the place in almost its primitive state. An army of carpenters and plumbers did wonders in a few months. They put in wood floors, five bathrooms, a garage for six cars, stables and a laboratory, and complete photographer's dark room with all the fixings for developing movies.

Green grass is almost nonexistent in Peking except in the legation compounds. Everything is brown loess clay. I determined to have a lawn in my gardens for the rock work and trees were very beautiful. So I purchased sod at an expense of about two hundred dollars. The grass was doing well except for a sprinkling of weeds which spoiled the velvet surface. Before going out to the Western Hills for a weekend, I showed a new gardener just how I wanted him to pull out the weeds by hand. He was said to be a good man on flowers but grass was something beyond his ken. When I returned on Monday, he proudly exhibited my expensive lawn. Every spear of grass had been pulled out by the roots!

But Chinese servants are really wonderful if you understand them and their ways. Each particular servant does a particular job and you must not expect him to do anything else; otherwise he loses "face."

You always have to think about "saving their face" no matter what they have done. One day my cook on his weekly bill had charged up eight dozen eggs, even though I had been alone in the house. The food was his particular "squeeze" and I was glad to ignore the fact that he kept ten per cent of what he had to spend, but I was unwilling to go further than that.

"Cook," I said, "it appears that you have used eight dozen eggs this week. I know that the meals you have been giving me are excellent and you, being a Number One cook, must use a lot of eggs to prepare such delicious food. I can't, however, afford so many eggs. Please give me bad food which can be prepared with only three dozen eggs a week in the future."

Of course my meals were better than ever because I had saved his face. There were eighteen or twenty servants about the place, and I believe their salaries totaled one hundred and seventy-five dollars a month (gold). They were supposed to feed themselves, but most of them didn't. My rice, flour, and tea went into their stomachs, but that was to be expected.

The Number One boy, who in a big household like mine was a real personage, hired and fired; I had nothing to do with it. He was responsible for the safety of all my possessions and if a servant stole anything, the Number One must make good the theft. He, in turn, could collect from the culprit's family. Thus, in order to protect himself, he never hired a servant who did not have a solid background.

About every two months the servants would begin to get a little slack in their duties. Then Lo would come to me.

"Master, more better you give everybody hell."

"All right, Lo, if you say so. What shall it be about?"

"Last time you give hell because courtyards not very clean. This time I think more better you say too many people in compound. Give very big hell. Everybody need it."

So when I could wipe the grin off my face and contract it into a suitable frown, I rushed into the servants' quarters like a whirlwind, yelling at the top of my lungs that I'd fire everyone in the house if they didn't keep people out of the compound, etc., etc. After it was all over, I called Lo.

"How did I do? Pretty good, don't you think?"

"Yes, Master do fine job. Only I think he going to laugh when he throw Number Two cook out of door. I almost laugh, too."

That had been a narrow squeak for me. In my simulated rage I had grabbed the first person I saw and hustled him to the gate, telling him to get out and not clutter up my compound. He was almost speechless with fright but managed to gasp: "But Master, I'm your cook."

And, by Jove, he was, only I'd never seen him. He was a student under my Number One who rated as about the best in Peking. Usually there were three or four "learn pidgin" cooks in the kitchen who did all the work. My Number One, who was the only one I paid, merely supervised and gave "diplomas" after a certain period under his direction. For this, he received their first month's wages after they had got a job.

After such an explosion, the service for a few weeks was priceless; then it would taper off and I'd have the job to do over again. But I didn't mind and I don't think the servants did either. It was custom.

As a change from Peking, many of us rented temples in the Western Hills or at the race course. Renting a temple sounds like a most sacrilegious performance but it's not like moving into a church. The Chinese were perfectly willing to let you spend the night in a temple or rent it year by year because you didn't interfere with their devotions.

My first temple rejoiced in the name of the Temple of the High Spirited Insects, and I took it with several friends. We were all sportsmen and soon became known as the Insects. We had an Insect polo team and an Insect racing stable; at point to point hunts, an Insect was always on the card and we swept the board at *gymkhanas*.

But the loveliest temple of all was one I had near the race course. It was called the Temple of the Hopeful Fecundity, and there women came to pray for children. In the outer court was a Number Two God and most of the suppliants went there first. If, after a suitable time, nothing happened, they had another go in the sanctum sanctorum which was where I lived. A huge gilded figure calmly sat on a lotus flower while two smaller ones kept guard on either side. Near them was a great bronze gong. If I happened to be "in residence" when a spot of business was to be transacted, the temple priest (who also acted as my caretaker) asked permission to bring in his client. She purchased a few sticks of incense, rang the gong, made her prayer, and went away completely undisturbed by my presence.

The temple was very, very old, about five hundred years, I think. The court was filled with twisted cedar trees, beautiful flowers, and singing birds which the priest kept in little bamboo cages hanging among the leaves. He was a gentle old man, and loved flowers passionately. Even in Peking my garden never could produce blossoms equal to those in the Temple of the Hopeful Fecundity. I suppose it was the result of constant and loving care. He personalized each plant and I used to hear him talking to them and calling them by name as he gently stirred the soil or arranged a support so that the flower would not break its stem.

I shall always remember him with admiration, and even a degree of affection, as a highlight in a country which had so many that I cannot remember them all.

Chapter 19

Marching Sands of the Gobi

On April 17, 1922, we left Peking for the first expedition to Mongolia. During the late winter our caravan of seventy-five camels had set out from Kalgan at the entrance to the great plateau, loaded with gasoline, food, and countless items of equipment. At last, the years of preparation, the months of strain and worry were ended. The field work had begun!

Certain incidents of those first exciting months in the Gobi stand out in my mind with photographic clearness.

One picture is of our motor cars, piled high with baggage, covered with brown tarpaulins, winding up the steep Wanshien Pass leading to the great plateau. Wonderful panoramas unfolded at every turn as we wound higher and higher. We looked back over a shadow-flecked badland basin, a chaos of ravines and gullies, to the purple mountains of the Shansi border. Above us loomed a rampart of basalt cliffs, crowned with the Great Wall of China, stretching its serpentine length along the broken rim of the plateau. Roaring like the prehistoric monsters we had come to seek, our cars gained the top of the last steep slope and passed through the narrow gateway in the Wall. Before us lay Mongolia, a land of painted deserts dancing in mirage; of limitless grassy plains and nameless snow-capped peaks; of untracked forests and roaring streams.

Mongolia, a land of mystery, of paradox and promise! The hills swept away in the far-flung graceful lines of a panorama so endless that we seemed to have reached the very summit of the earth. Never could there be a more satisfying entrance to a new country.

I remember, particularly, our second camp in the grass lands. The first night had been spent in mud with mired cars and was filled with strenuous work. But the next evening we camped early in an amphitheater of rounded hills. I had shot an antelope and a bustard. For the first time, we gathered round a camel-dung fire on a perfect windless night, ate and smoked and talked while the stars came out and a lone wolf howled on a hilltop to the west. All of us were friends with mutual trust, affection, and respect. We had entered upon an expedition which our colleagues said was a waste of time. We knew it was a scientific gamble; that we were on the threshold of a great adventure which would make all our reputations if we were right and ruin mine if our reasoning was wrong.

In those moments, we were very close together, telling unashamed our hopes and fears and evaluating the chances of success. Every man sitting about the flickering little fire had been selected because I believed him to be the best in the world for that particular job, not only because of his scientific knowledge, but also because of his personality. In the desert, we had to create our own little world. There were no newspapers, letters, telegrams, or other diversions. Whatever we had, we had to make for ourselves. Day after day, we saw the same faces, learned every mannerism, the innermost secrets of a man's character.

When one is at grips with crude nature, when the struggle to maintain oneself and do one's work against physical odds, is really on, the gloss of civilization quickly fades. It leaves character bare for all to see. There is little left to know about a man after one has lived with him in the field for a few weeks under primitive conditions. I had brought together a group of men who were considerate of their fellows, generous, unselfish, ready to accept the worst with the best; men so keen on their jobs that hardships were incidents; men with such steadfastness of purpose that nothing could turn them aside.

At least this was the way I had planned it, and when I went to sleep that night I knew I had those men. Nor did I ever change my mind. During all the years we were in the desert together, not one of them ever let me down.

I shall always remember the day we found the first fossils. Before we went to Mongolia, only one fragment of a "rhinoceros" jaw had been discovered on the whole Central Asian plateau. At a promising looking exposure of yellow gravel on the edge of a great basin filled with camel sage, I dropped Drs. Berkey, Granger, and Morris.

"You stop here and have a look," I said. "I'll make camp five miles away where the caravan should have left us a dump of gasoline."

The tents were pitched at the base of some gray-white ridges. While I was watching a sunset, which splashed the sky with gold and red, the geologists' cars roared into camp. I knew something unusual had happened for no one said a word, but Walter Granger's eyes were shining. Silently he dug into his pocket and produced a handful of bone fragments; out of his shirt came a titanothere's tooth and the various folds of his clothes yielded other fossils. Berkey and Morris were loaded in a like manner. Walter stuck out his hand.

"Well, Roy, we've done it. The stuff is here. We picked up fifty pounds of bone in an hour."

Then we all laughed and shouted and shook hands and pounded one another on the back and did the things men do when they are very happy. No prospector ever examined the washings of a gold pan with greater interest than that with which we handled that little heap of fossil bones. Some we knew were rhinoceros and we were sure that others were titanotheres. That was what was so exciting, for *no titanotheres had ever before been discovered outside of America*. The other specimens of smaller animals were not positively identifiable.

While dinner was being cooked, Granger wandered off along the gray-white outcrop that lay like a recumbent reptile west of camp. Even in the failing light he found half a dozen fossil bones. We realized that we had another deposit at our very door. The next morning, just as I was starting out, Dr. Berkey returned to camp. For a distinguished professor of Columbia University, he was acting very queerly, but he would give me no information. "Come with me," was all he said.

When we reached the summit of the outcrop, I saw Granger on his knees working at something with a camel's-hair brush.

"Take a look at that and see what you make of it," said Berkey. I saw a great bone, beautifully preserved, outlined in the rock. There was no doubt, it was *dinosaur.*

"It means," said Dr. Berkey, "that we are standing on Cretaceous strata of the Age of Reptiles—*the first Cretaceous strata, and the first dinosaur ever discovered in Asia north of the Himalaya Mountains."*

Unless one is a scientist, it is difficult to appreciate the importance of that discovery. It meant that we had added an entirely new geological period to the knowledge of the continental structure of Central Asia and had opened up a palaeontological vista of dazzling brilliance. With the rhinoceros and titanothere teeth, the dinosaur leg bone was the first indication that the theory upon which we had organized the expedition might be true; that Asia is the mother of the life of Europe and America!

A few days later, we met our caravan. The rendezvous was to be at Tuerin, the base of an ancient mountain, ages ago of majestic height but now reduced by wind and weather to a ragged mass of granite rising like a time-worn citadel nearly a thousand feet above the plain. We came to the base of the "mountain" at noon and in the distance saw a great caravan camped beside the trail. Soon, I made out the American flag streaming from one of the loads and recognized our boxes. Merin, the leader, said they had arrived only an hour before we came. This was the day on which I had told Merin, months earlier, to reach Tuerin.

We camped in the center of the mountain, on a grassy plain with orders for the caravan to follow. I climbed to a flat-topped ledge just as the great white leading camel, bearing the American flag, appeared from behind a boulder in the rocky gateway. Majestically, in single file, the animals advanced among the rocks and strung out in a seemingly endless line. My blood thrilled at the sight. The camels swung past the tents, broke into three files like soldiers, and knelt to have their loads removed; then with the usual screams and protests, they scrambled to their feet and wandered down the hill slope to the plain, nibbling at the vegetation as they went.

In the intervals of repacking the caravan loads, we explored the innermost recesses of the jagged peaks near camp. Everywhere the piles of empty rifle shells, cartridge clips, and partially eaten human bodies gave evidence of battle. The past winter had been one of tragedy for Mongolia. In 1920, "Little Hsu," a Chinese general, had taken possession of Urga by a clever trick. But Baron Unger-Sternberg, a fanatic White Russian, had made a secret bargain with the Mongols. In return for a base from which to operate against the hated Bolsheviks, he agreed to rid Mongolia of Chinese soldiers. The battle at Urga had been short and bloody. In two days the Chinese were fleeing for their lives. At Tuerin, where we were camped, four thousand Chinese soldiers peacefully enjoyed snug quarters about their *argul* fires. The "Mad Baron" sent Cossacks to attack them, but before the Russians arrived a Mongol general, by hard riding, reached Tuerin at the head of three hundred men. "We got there at dawn," he told me later in Urga, "and I gave the men ten minutes. Then we attacked. We rode full speed through the camp, killing everyone we saw. Then we rode back again. The Chinese ran like sheep and we butchered everyone."

Except for the modern weapons, the story might have been a thousand years old, for this method of warfare was a heritage from Genghis Khan. Hours of hard riding, regardless of sleep or food, a sudden whirlwind attack and then relentless slaughter!

But the Mad Baron's tenure of Mongolia was short-lived. He was defeated by the Russians. Two of my American friends, who spent that winter in Urga, told me of the horrors of the Red advance. They saw men, women, and children butchered in the streets of

Urga or swinging by the neck from their own door posts, dogs gnawing at frozen bodies in every alley, and sights of horror which sent them home with shattered nerves.

Such was the background, only a few months past, of our entry to Mongolia. On my first visit it was a land of wild, free tribesmen, a frontier country like that of our own West in 1840. But those days were past. When we went there in 1922, bands of merciless Bolsheviks were hunting down the White Russians on all the caravan trails and at every well. Chinese and Mongol brigands skulked like jackals on their heels, gleaning what grisly fragments they could find from the leavings of the Red marauders.

Never will I forget one night when we camped near a stream on a velvety lawn of green grass.

"Why are there no yurts here," I said to my Mongols, "and sheep and goats?"

The men were puzzled. They shook their heads.

"There must be some reason. We don't like it, we Mongols. Never would this place be without a dozen yurts if all were well."

The answer came within an hour. Skulking on a ridge above the stream our men found two wretched natives. Half demented, trembling with fear, they were brought into camp. After food and tea they told us their tragic story.

"We had ridden to a village a hundred miles away to buy two horses," they said. "Twenty yurts were here when we left; our fathers and mothers and all our families. But a Chinese caravan from Uliassutai laden with rich furs stopped to rest and feed their ponies. During the night a band of Bolsheviks rode in. They murdered every soul, including all the Mongols, and drove off the camels with the furs. They threw the bodies in the gully over there, among the rocks. We came back just in time to see it from the ridge where you found us. It was only a few weeks ago. Since then we have lived in a little cave. We don't know what to do."

Sure enough, in the deep ravine among the heap of rocks lay a mass of human bodies, already half eaten by the wolves and ravens.

For more than a month in the early spring sandstorms made life miserable. An entry in my journal on June 14 tells of what we had to face day after day until July brought heat and mirage:

There had been a gentle breeze, but suddenly it dropped to a dead calm; a heavy stillness, vaguely depressing. Slowly I became conscious that the air was vibrating to a continuous even roar, louder every second. Then I understood. One of the terrible desert storms was on the way. The shallow basin seemed to be smoking like the crater of a volcano. Yellow "wind devils" eddied up and swirled across the plain. To the north an ominous tawny bank advanced at race horse speed. I started back toward camp, but almost instantly a thousand shrieking storm demons were pelting my face with sand and gravel. Breathing was difficult; seeing impossible. I stumbled over the rim of the basin, and tried to strike diagonally toward the tents. Even the ground beneath my feet was invisible. In perhaps ten minutes, perhaps half an hour, I stumbled into a depression and lay there trying to think.

Suddenly forms took shape in the smother right beside me. I reached out and caught one of them by the leg. They were three of my men. Pressing our mouths close against one another's ears we held a consultation. One thought the tents were directly south of us. I had no idea where they were. Clinging together we groped our way through the blinding murk. At last, we stumbled against a black object. It was the cook tent, still standing but in danger of being torn to shreds with every blast. We felt our way inside and lay on the ground, faces buried in wet cloths; it was the only way we could breathe.

The gale continued for an hour and then suddenly dropped into a flat calm; not a breath stirred the American flag which hung limply above my tent, whipped almost to ribbons. The silence was uncanny after the roar and rush of the storm. Just as we were crawling out of the tent we heard a shout and saw a brown figure coming into camp. Behind the broad grin on the desert-colored face was Walter Granger. When the storm broke, he had fought his way to a partly excavated titanothere skull, to mark the spot for fear it would be lost in the sand. He reached it and huddled into the pit with his face in a coat. He had been completely buried, except for his head, and well nigh smothered.

We dug out the tents and emptied the sand from our clothes and beds. Half the Gobi seemed to be in our belongings and it had sifted even into the tightest boxes. I sent a car to the Well of the Mountain Waters a mile away, and everyone had a bath and dressed in clean clothes. We felt human once more. But while dinner was being served one of the men looked toward the north and gave a shout. There it was again—the same tawny cloud! This time it was preceded by an enormous "wind devil" which danced wildly across the plain. It was heading toward us and we knew what to expect if it struck our camp. I called all hands to weight the bottoms of the tents and pound in the pegs. Explosions of wrath came from every man because we were so clean and knew, full well, how dirty we would be in a moment.

The attack came with a crash and a blast of gravel like exploding shrapnel. For five minutes the sand spout whirled round the camp as if trying to suck the tents and all our belongings up into the smoking vortex. Then, repulsed at every point, it danced away across the plain and vanished in the distance.

The wind began again and developed into a full gale before the hour had passed. For ten days it blew without a rest until our nerves were worn and frayed. But we knew that sometimes it would end and give place to the perfect sunlit days that make Mongolia in the summer a land of joy.

The evening of August 4 was one of the high spots of the first expedition. We were camped on the shore of a lovely desert lake called by the Mongols, Chagan Nor, the "White Lake." Across the water, beyond a rim of fantastic sand dunes, toward the eastern Altai Mountains lay a new and enticing world, completely unexplored.

My journal says:

Late in the afternoon, there was a little rain and just at sunset a glorious rainbow stretched its fairy arch from the plain across the lake to the summit of Baga Bogdo. Below it the sky was ablaze with ragged tongues of flame; in the west billowy gold-margined clouds shot through with red, lay thick upon the desert. Wave after wave of light flooded the mountain across the lake—lavender, green and deepest purple—colors which blazed and faded almost before they could be named. We exclaimed breathlessly at first and then grew silent with awe. Never might we see the like again. Suddenly a black car with Granger and Shackelford in it came out of the north and slipped quietly into camp. Even Shackelford's buoyant spirit was stilled by the grandeur of what was passing in the sky. Not until the purple twilight had settled over mountain, lake, and desert, did the two men tell us why they had been so late. They had discovered parts of the skeleton of a *Baluchitherium!*

The "Beast of Baluchistan" was the largest mammal ever known to have lived upon the earth. Only a neck vertebrae and foot bones had been found in India by my old friend C. Forster Cooper of Cambridge University, England, and one could only guess what sort of creature it would prove to be. Now the mystery would be dispelled. Except for the remains of primitive man, no discovery could have been more exciting to a palaeontologist.
My journal goes on to say:

The parts Granger had discovered were the end of the humerus, or upper fore-leg bone, and the whole side of the lower jaw with the teeth as large as apples, well preserved. I went to sleep very late that night, my mind full of *Baluchitherium* and had a vivid dream of finding the creature's skull in a canyon about fifteen miles from the spot where the jaw had been discovered the day before. I determined to go back to the place and make a further search.

With Shackelford, and a Chinese chauffeur, Wang, we returned to "Wild Ass Camp" the next afternoon. The two other men set to work in the bottom of the gully while I inspected the side, now and then sticking my pick into a bit of discolored earth. From the summit of the tiny ridge, I looked down the other side. Fragments of bone peeped out of the sand in the bottom of the wash. Its color was unmistakable. With a yell, I leaped down the steep slope. When "Shack" and Wang came around the corner on the run, I was on my knees, scratching like a terrier. Already a huge chunk of bone had been unearthed and a dozen other bits were visible in the sand. They were so hard we had no fear of breaking them. Laughing in hysterical excitement, we made the sand fly as we took out piece after piece of bone.

Suddenly my fingers struck a huge block. Shack followed it down and found the other end; then he produced a tooth. My dream had come true! We had discovered the skull of a *Baluchitherium*.

At six o'clock, while the men were having tea, we burst into camp, shouting like children. Granger is not easily stirred but our story brought him up standing. He was as excited as the rest of us. Even though we realized the "Baluch" was a colossal beast, the size of the bones left us astounded.

When the remains were later studied in the Museum by Professor Osborn, he confirmed the original supposition that it was a giant, hornless rhinoceros and the largest known land mammal. It reached a height of seventeen feet at the shoulders, was twenty-four feet in body length, had a long neck, stilted limbs, and probably prehensile lips adapted to feeding on the herbage of the higher tree branches like a giraffe. The "Baluch" lived during the Oligocene Period, about thirty-five or forty million years ago, and was so highly specialized that when the climate and vegetation changed, it became extinct without ever leaving Central Asia.

September first was the greatest day of all for the Central Asiatic expedition. We were almost in the center of the Gobi. I was eager to leave, for geese and ducks, flying southward, sand grouse gathering in flocks of countless thousands, and golden plover arriving from the northern tundras warned that winter might drop on us suddenly at any moment. The cold didn't worry me, but every day increased the danger of blizzards which made traveling by car difficult and dangerous. I had had one experience of Mongolian blizzards and didn't want another.

The Russian maps, the only ones in existence, showed a great blank space between us and the Sair-Usu trail where we were to rendezvous with the caravan. Nothing was marked on the unexplored area except a range of mountains six or seven thousand feet high. I laid a compass course directly across the mountains. How the motors could negotiate them I didn't know. Perhaps we could find a pass or a break in the chain.

For three days, we ploughed along over rough going, without sight of even a hill. I was getting worried for those mountains were very much on my mind. On the third night, Berkey and Morris computed our position from star sights. We were far beyond where the range was marked upon the map. The mountains simply didn't exist. Not a Mongol had we seen for more than a hundred miles. Water was very low. At last three yurts showed in the distance and the fleet halted while I drove over to inquire about a well.

I had thought that that was all there was to this stop, but Shackelford, whose inquisitive mind was always busy, noticed some peculiar looking stones on the plain and walked over to investigate. From there the rim of a great red basin was just visible. He decided to spend five minutes looking for fossils before returning to the cars. Halfway down the steep slope, a white skull, about eight inches long, rested on the summit of a sandstone pinnacle. He picked it off and hurried back. None of us had ever seen its like. Granger was only able to say that, without doubt, it was a new type of reptile, unknown to science. The Mongol, whom I had found, told us there was a well in the basin floor. Obviously we must camp there for the night.

The badlands were almost paved with white fossil bones and all represented animals unknown to any of us. Granger picked up a few bits of fossil egg shell which he thought were from some long-extinct birds. No one suspected, then, that these were the first dinosaur eggs ever to be discovered by modern man—or to be identified. Neither did we dream that the great basin with its beautiful sculptured ramparts would prove the most important locality in the world from a palaeontological standpoint. In the late afternoon sun the brilliant red sandstone seemed to shoot out tongues of fire and so we named the spot the Flaming Cliffs.

I shall always remember our return to Kalgan. We drove through the narrow streets of the Chinese frontier town, our cars covered with the yellow dust of the Gobi, with cut-outs open and blaring horns. From the doors of every shop people rushed out, lined the streets, and cheered to welcome us home. They had expected never again to see us when we had driven away five months before.

At the British-American Tobacco Company's mess we bravely tried to live up to the requirements of civilization. They gave a dinner party that night for all Kalgan residents. Each one of us had some article of adornment that he had cherished for the homecoming. Shackelford appeared in a wonderful blue shirt. I had a purple necktie, and Bayard Colgate and Walter Granger each produced a pair of new shoes. Yet when we assembled at dinner we all felt strange and uncomfortable.

Walter and I tried to sleep on a soft bed in a stuffy room. We tossed and turned and smoked cigarette after cigarette. Sleep wouldn't come. Finally, Walter said: "Hell, Roy, it's this damned bed. Let's get our sleeping bags and bunk outside."

We stole out like thieves, unrolled our fur bags on the earth of a defunct flower bed, looked up at the stars for a few minutes while the soft wind caressed our faces, and dropped into a dreamless sleep.

Chapter 20

The Emperor's Bride

The day after arriving in Peking the whole staff gathered in the drawing room of the headquarters on the "Bowstring Street." We were dressed and shaved and had taken on the restraints of city life along with its habiliments. The men sat on the silk-covered sofas beside the fire and looked at each other almost as though they were strangers. No one said anything. It was pretty awful. Something had to be done. I called my Number One boy: "Lo," said I, "cocktails, and make 'em strong. It's only eleven in the morning, but what of it?"

We toasted each other and the success of the expedition solemnly. Lo, unobtrusively, saw that every glass was filled as soon as it was emptied and in fifteen minutes the tension had broken.

"Now," I said, "we must compose a cable for the Museum. This will be the first authoritative statement of the results of the expedition. What the foreign correspondents have sent out already is only hearsay. Let's make it good."

For two hours we labored over that message and produced a statement of the most important results of the expedition; when we got through the cable was something of a masterpiece. It told everything I was sure the newspapers would want to know—or at least ought to want to know. We had made scientific history and I wanted it to get into the public record as soon as possible.

After all, this was the first time that motor transport had proved successful in extended scientific exploration. Our cars had covered three thousand miles and all were fit for next year's work. Knowledge of this achievement was important not only to science, but to commerce as well, for it demonstrated that remotest Mongolia and, therefore, many of the earth's little known regions were accessible to motor transportation.

Scientific results had surpassed our greatest hopes, yet we knew that we had only scratched the surface. The expedition had traveled from Urga by way of Sain Noin Khan to the Altai Mountains, where vast fields rich in Cretaceous and Tertiary fossils had been discovered, including the fine skull and parts of the skeleton of *Baluchitherium,* the largest known land mammal.

We had also obtained complete skeletons of small dinosaurs and parts of fifty-foot dinosaurs; skulls of rhinoceros; hundreds of specimens, including skulls, jaws, and fragments, of mastodon, rodents, carnivores, horses, insectivores, deer, giant ostrich, and egg fragments. We had found wonderfully preserved Cretaceous mosquitoes, butterflies, and fish, unknown reptiles, titanotheroids, and other mammals.

The geologists had identified extensive deposits of Devonian Carboniferous and Permian age Palaeozoic rocks. They had measured twenty thousand feet across the upturned edges of Jurassic strata. An enormous granite bathylith had been discovered comparable to the Laurentian bathylith of Canada, with a wonderful development of roof pendants and contact metamorphism. We had the longest detailed topographic route map and continuous geologic section ever made on reconnaissance. We had mapped a thousand-square-mile strip in the type region of Mongolian geology.

Our photographer had obtained twenty thousand feet of film of all details of the work of the expedition, and feature films of every phase of Mongol life. He had filmed large herds of antelope and wild ass, never before photographed alive. He had also obtained five hundred still photographs.

The zoological work had been extraordinarily successful. We had secured the largest single collection of mammals ever taken from Central Asia, including many new species, and materials for fine habitat groups of the wild ass, antelope, ibex, and mountain sheep for the Museum hall of Asiatic life.

Clifford Pope working in North China and the Ordos desert had obtained a splendid collection of fish and reptiles. He had made a comprehensive zoological survey of Shansi Province.

All of that we put into the cable, and just reading it over made us realize, as we scarcely had before, what an expedition it had been. At once everyone began discussing plans for extensive field work the following summer.

After the initial cable, we had sessions with the foreign correspondents, all of whom were our personal friends. Congratulatory messages poured in. I must say that those scientists who had been loudest in their prediction of failure were first in admitting that they had been wrong.

Professor Osborn cabled: "You have written a new chapter in the history of life upon the earth." Messages came from scientific and geographical societies in all parts of the world: America, England, Australia, France, Germany, Hungary, and Sweden. It was satisfying, to say the least. Moreover, we had had an experience that never again could come to any living man for the Central Asian plateau was the last great region scientifically unknown. Almost everything was new. The mammals, reptiles, fish from the desert lakes, the geology, and particularly the fossils wrote illuminating words upon pages in the record of natural history which had hitherto been blank.

Immediately, we discovered that our explorations were of commercial, as well as scientific, importance. All unwittingly we had opened Mongolia for motor transportation. Representatives of Chinese importing firms asked how they could get cars to various points in the Gobi to bring out valuable furs; contract for hides, camels, and sheep's wool and ponies; what routes to take; where to send gasoline and a dozen other questions. By the end of our second expedition a score of cars were using the trails we had mapped. It was a striking example of how quickly commerce follows on the heels of exploration.

Since the first expedition was only a reconnaissance to explore and appraise the country, we prepared for a second year of intensive work in the localities already discovered, with a larger staff. Some of the men returned to America for the winter; others remained to carry on explorations in various parts of China. I settled down to a winter in which both work and play were mixed.

There was much to do. We would have a staff of forty men, all told, and I must buy more camels, get additional equipment, food, and other supplies which the first year had shown to be necessary for the comfort and health of the expedition members. Everything had to be thought of in advance, from shoe strings to chewing gum. The desert gave us nothing but game.

But I was not too busy to throw myself into the social and outdoor life of Peking with enthusiasm. After all, why not? There was behind me the satisfaction of a job well done and the prospect of another summer of exciting work in the Gobi. I didn't neglect my work, but I did have a good time. The autumn polo tournaments had just begun. Three or four hours every day I spent on horseback, and when polo ended in November, point-to-point hunting and a drag with hounds began. There were ponies to train for racing and to school in jumping every day. I bought the best and rode them hard.

One could get a good Mongol pony for from fifty to a hundred dollars, but it was much like buying a pig in a poke. Perhaps the animal might be a washout, or a winner at polo, hunting, or racing. Usually my stable held ten or twelve ponies. I got to be pretty good at judging the merits of these extraordinary little beasts and seldom lost money for they could always be sold as hacks. David Harum had nothing on me when it came to horse trading.

I once bought a pony from Major Magruder, who is now a brigadier general and was head of President Roosevelt's military mission to China. Squire was the pony's name. He was a magnificent animal but a bad actor. Anyone who got on his back took one ride and that ended the relationship. John Magruder sold me Squire for a hundred dollars with the warning that I was being "done in the eye" because he never would be usable. But I had a theory that every time Squire bolted it was because they bitted him more and more severely; that really he had a very tender mouth and should have the opposite treatment. I took him into the country where there was plenty of room and tried him with a rubber bit, as gentle as anything could be.

Squire acted like a lamb. He became as devoted as a dog and would follow me all over the compound. Once, on a bet, I walked into the laboratory around half a dozen tables and out the door, with the pony right at my heels. He had a system of signals with his ears which I understood perfectly. When everything was all right and he had settled down to hard running, both ears were laid straight back. If he saw an unusual jump but one which he understood, one ear went forward and one back. But if he wanted assurance, both ears stuck straight out. Then I knew he would wait for me to give him the signal to take off with my legs.

We won every hunt in which I rode him. He was as jealous as a cat and wouldn't let me touch another pony, or even my dog Wolf. No one else could get on his back. The only exception was the late Kermit Roosevelt, who visited in Peking. Kermit was very good with horses. By dint of much coaxing by me and explanation that Kermit was a friend and I wanted him to have a good ride, Squire let him mount. The gallant little pony won the hunt over thirty jumps even though his rider weighed two hundred pounds.

Finally, when I had to go to America for an unusually long stay I sold him to a friend in Shanghai for three thousand dollars, with the understanding that I could buy him back at the same price upon my return. I traveled much of the time with him in the box car to Shanghai and took the bit and saddle he liked. I hated Shanghai but spent a week with Squire introducing him to his new owner. When my ship sailed I had to say good-by and I am not ashamed to admit that my eyes were filled with tears. He was only a Mongol pony but he loved me with all his great heart and I loved him. I never saw Squire again. He won the Far East Grand National and ten thousand dollars for his owner. Three months before I returned he broke his back on a slippery jump in training.

Polo was a passion with me. By dint of hard practice I got to be reasonably good and always rode as one of the four which represented Peking in the Interport matches between Shanghai, Tientsin, Hankow, and regimental teams from all over the East. The polo matches were brilliant affairs with gay crowds, bands, and military trappings. I don't think I ever had more excitement than when I waited, every muscle tense, for the referee's whistle which started play. Once we were in action, of course, all my nervousness vanished, but those first tense moments were something to live for.

One of the most amusing characterizations I ever heard of myself was at an important match between the U. S. Fifteenth Infantry of Tientsin and our Peking four. The captain of our team was the late Colonel Margetts, military attaché of the American legation, who was an old West Point rival of Colonel Taylor of the Fifteenth. The officers of the regiment had come up almost *en masse* determined to win the match. They had a brilliant player, Lieutenant Cornog, at back, and as I was number one on our four, he was my especial assignment. Colonel Margetts said to me: "I don't care if you never hit the ball but *keep Cornog out of the game*. Ride him like hell every minute and we'll take care of the others."

I had fast ponies and was much lighter than Cornog. During the match which we won, eight to nothing, I devilled him every second. At the end we were sitting in front of the pavilion having a Scotch and soda when I heard Colonel Taylor say to Cornog in a very doleful voice: "Lieutenant, I was much disappointed in your game today. You didn't seem to get in it at all. Why, I don't remember that you ever hit the ball. What was the matter?"

"Well, Colonel," said Cornog, "how in hell could I get into the game with that bald-headed bone-digger sitting in my lap all the time?"

I overheard the remark and roared with laughter, much to Cornog's embarrassment.

"Don't you mind, Jug," I said. "That story will be worth a thousand dollars to me." And it was.

Like most places in the Far East, Peking residents took their sport very seriously. Apparently it was much more important than business and bitter feuds developed, mostly over ponies. A dozen or more people who were continually appearing at the same functions in our small circle wouldn't speak to each other. It was a silly situation and one which tickled the Puckish sense of humor of one of my friends. He was about to leave for Europe after three years' residence in Peking. I, also, was off for the Gobi within a week.

"Roy," he said, "let's you and me give a dinner at my house at which we'll have only those people who have hates against each other. No one else; only the feudists. We are both going away and we don't care what anyone says about us. It ought to be a lot of fun."

I thought it a grand idea and the invitations went out by special "chit-coolies" so that no one knew who else was coming. The night of the dinner, my friend and I stood in his great drawing room to welcome the guests. As they came in, the expressions on their faces were excruciatingly funny. Most of them acted as though a skunk had suddenly been let loose in the room. By the time the twelve or fifteen guests were assembled, some of them had begun to smell a mouse. Our special cocktails were awfully potent, however, and they softened up the company considerably before dinner was announced. We had arranged the place cards so that every man or woman who was not on speaking terms sat next to each other. The first courses were pretty grim with conversation going on triangularly across the table but none with the next door neighbor. My friend had some superb champagne which was famous throughout Peking. When it arrived he rose and said: "As you may have suspected, this dinner was a put up job. Roy and I decided we'd do the only missionary work we've ever done by trying to patch up some of these silly quarrels. We're both leaving next week, so we don't give a damn what you think of us. This is the last case of my best champagne. Suppose you all stand up, drink to the happy departure of your hosts, and everyone shakes hands with the person he or she dislikes most."

For a moment there was dead silence and no one moved. Then one of the men turned to his neighbor with a grin and stuck out his hand. That was the signal for a round robin and the rest of the evening went off in a gale of laughter. I don't think all the quarrels stayed patched up but I know some of them did. Much to our surprise, we were praised instead of damned. We didn't deserve to be, for in the beginning we had conceived the party purely in the spirit of devilment with no intention of doing a good deed. That part of it came incidentally after the evening was well started and it was evident that there were potentialities for a boy scout act.

During that winter of 1923 came the marriage of the Boy Emperor to the daughter of a Manchu nobleman. I was present by grace of the American minister for only a few diplomats were invited. Would that a great writer had been there—a man with the gift of words, that he might adequately describe what I saw that moonlit winter's night; that he might make you feel the deep and sad significance, the solemnity and the hopeless futility that hung like a pall over those few remaining nobles of the Manchu court! Ostensibly it was a wedding, but more truly did it signify the burial of the hopes of a great race which for three hundred years had sat upon the Dragon Throne.

The little bride lived in a big house in East Peking. The emperor had been relegated to the northern part of the Forbidden City. The Chinese grudgingly admitted that the emperor could not be denied a wedding with some elements of royal precedent. But they gave what they felt they had to give with bad grace and subtle disdain. It caused the Emperor to lose face and must have been a bitter pill for the Manchus to swallow. They would not allow the bride to be admitted through any of the main gates which led directly to the throne room. Instead, she had to come by the north gate, the back door of the Forbidden City. They could not, however, deny her the traditional yellow path. All the afternoon carters with heaps of fine golden sand sprinkled the road from her home to the emperor's palace.

It was, I remember, the night of the St. Andrew's Ball, and Peking Scotchmen were hosts to the foreign community in the new hotel. All was dancing and merriment. At supper a sergeant of the Black Watch bore the ceremonial haggis above his head escorted by pipers in kilts and tartans. At half past two the American minister signaled to me from across the table. The mercury stood at ten degrees below zero in a brilliant moonlit winter's night. Bundled in great fur coats and foxskin caps we drove up the Nan Chi-tzu, beside the palace moat, overlooked by the golden pavilions of the Forbidden City. There, until 1911, the emperor's guard had stood on watch day and night for three hundred years. Now the ramparts were deserted. The car swung sharply around the corner of the wall to the north gate. As a crowning indignity, the Manchu banner-men had been withdrawn and black clad Chinese police inspected all the passes. But inside the gate, eunuchs dressed in ceremonial robes ushered us through half a dozen courtyards flooded with moonlight to a great waiting room, obviously a storehouse. The only decorations were a dozen ornate hanging lanterns. The place was filled to overflowing with Manchu nobles in full court dress, even including peacock plumes for those of lesser rank. A young duke, keeper of the imperial graves at Tungling, signaled a eunuch to bring us tea. The place was deathly cold; the airless, ice-box cold that bites into one's bones through the heaviest clothes. I shivered in my thin dress suit and pulled the fur coat tighter but it could do nothing for my freezing feet. The Manchu noblemen were clothed in sable robes and fur-lined boots like those one hoped to find, but seldom did, in the silk shops beyond the Chien Men Gate.

We spoke gravely with our hosts in the stilted phrases of court language. All were Manchus, members of the nobility who, for some reason known only to themselves and the Chinese, had escaped death in the first bloody days of the revolution. There was no laughter or light talk as in most Chinese gatherings. A solemnity and sadness hung like a black cloud over the room. Never again, they knew, would a Manchu sit upon the throne of the celestial kingdom even though the infant republic did not survive. If there were another emperor he would be Chinese. Their day was done.

I drew off in a corner by myself and looked at the men gathered in that bleak room so different from the place where the last emperor had received his bride. Fine faces, most of them. All were middle aged or old; only four or five younger men. They looked drawn and haggard in the yellow candlelight and very hopeless. But a grave serenity in their greetings made you feel that if death came to them on the morrow, as well it might, they would meet it with the dignity of great gentlemen.

Suddenly a bugle sounded. The Manchus moved unhurriedly into the court to range themselves on either side of the marble causeway where only the emperor might walk. I heard sharp commands, the clatter of rifles, and the huge red gates swung wide. Framed in the opening, carried by eight bearers in embroidered robes, was the golden Phoenix Chair in which for three centuries the royal brides of China had come to the Forbidden City. For an instant they paused, then moved swiftly through the lacquered doorway. Without warning, as one man, the Manchus dropped upon their knees, foreheads touching the ground. Automatically I followed suit. Only the ministers remained partially erect, bent over in deep bows. I, for one, could have done no less. It was all over in a few seconds. The chair passed between the kneeling throng, through a yellow carved gate into an inner court, and was lost to sight.

I got to my feet, too deeply stirred to speak. The American minister also was profoundly affected. It had been so spontaneous, so reverent, and so dramatic that I felt we had lived through one of the great moments of history. Indeed it was. I had witnessed the last act in the age old drama of imperial China.

What followed was an anticlimax. No foreigners witnessed the actual marriage ceremonies, the tea drinking, and the exchange of gifts. A chosen few, however, were admitted to an afternoon reception. The emperor, his weak eyes hidden behind dark spectacles, stood beside his bride, impassive, unsmiling. She was lovely. A little heart-shaped face, soft brown eyes, and a body slender as a willow wand. Her hair was done in the traditional style of a bride whose marriage has been consummated. The emperor spoke English fairly well, for a scholarly British diplomat had been his tutor, but he was shy of speech. I offered my congratulations in Mandarin Chinese. He answered in the same language.

A few years later, when the missionary-exalted "Christian General" Feng Yu-hsiang controlled Peking and systematically looted the imperial treasures, there was a rumor that the emperor might be killed. The Japanese saw a heaven-sent opportunity. Trying to conceal their participation in the plot, yet fooling no one, they engineered his escape to Tientsin. But "sanctuary" in the Japanese concession turned out to be as much imprisonment as that which the emperor had known in the Forbidden City. They kept him "on ice" until the political stage was set to invade Manchuria in 1931. Then, in due course, they announced he would be crowned "Emperor of China" in the homeland of his dynasty.

Of course, the hope was that the Chinese, disgusted with the chaos of the infant republic, would rally to his standard, start a new revolution, and bring him back in triumph to the Forbidden City. Through him the Japanese would then control North China. But the deep laid plans went sadly awry. The Chinese refused to take the bait.

I had an audience with the emperor, in Manchuria, in 1932. He was a pathetic puppet, surrounded by hissing Japanese; an ineffectual boy caught up in the maelstrom of a political flood which whirled him about like a straw upon the waters. Even had he been wise enough to know what it was all about, doubtless the result would have been the same. He was born of an imperial line and as such his body had a potential value to those who held it. Poor little emperor!

Chapter 21

Where the Dinosaur
Laid Her Eggs

Exactly a year after our first expedition (April 17) we left Peking
again for the Gobi. For a month there had been an unprecedented
number of robberies along the camel trail north of Kalgan and I
was a little worried about our caravan. When we arrived at the
rendezvous in the desert the camels were not there. Neither were
there any reports of them from Mongols who had traveled the trail
which it should have followed.

After waiting a week, I was just on the point of going out on
horseback with five or six men who were spoiling for a fight, to
scout the country, when Merin, the leader, rode into camp on his
big white camel. The caravan was close behind. The Mongols
arrived, gleeful as children to be safe in camp. Hearing that there
was a band of five hundred brigands ahead of him, Merin had
slipped off into the desert. He traveled only at night, from well to
well, and camped during the day in sheltered hollows where he
could not easily be seen. His weather-tanned face simply beamed
as he told how he had played hide-and-seek with the bandits and
yet had filled the stomachs of his camels with some of the best
grazing they had had all winter.

A short time later, on the way back to Kalgan for some extra
supplies, I had an amusing experience with brigands. We had two
cars; I drove one and Young the other. Mine was a couple of miles
in advance when I came to a deep valley where two Russian cars

had been robbed only a week earlier. The bandits had taken twenty thousand dollars' worth of sable skins and killed one man. The other chap had been stripped absolutely naked and left to find his way to Kalgan.

Just before reaching the spot, I saw the head and shoulders of a man on horseback showing over the summit of the hill, three hundred yards away. The sun glinted on a rifle barrel. Now, there were only two kinds of men who carried rifles in China—bandits and soldiers—and, at that time, the two were synonymous. Anyway, I had no mind to have him there, whoever he was. I dropped a bullet from my .38 revolver too close for comfort but didn't try to hit him. He disappeared abruptly.

Just then my car swept over the rim of the basin and started down the steep slope. In the bottom two hundred yards away were four horsemen, rifles on their backs. I knew instantly they were bandits and I was in for it. The trail was narrow and rocky and I couldn't turn; also I knew that a Mongol pony never would stand against the charge of a motor car. Opening the cutout, I stepped on the accelerator and the car rushed down the hill roaring like an airplane. The ponies went mad with fright. At first the bandits tried to get the rifles off their backs, but in a moment their chief concern was to stay in their saddles. Three of the ponies rushed wildly across the valley, rearing and plunging madly. The fourth seemed too frightened to run. I was right beside him and I'll never forget the look of abject terror on the face of that Chinese brigand!

The revolver was in my right hand and, of course, I could have killed him easily but there was no sense in doing that. But the peaked Mongol hat he wore bobbed up and down and was too great a temptation to be resisted. I fired at it four or five times, trying to knock it off his head. Finally his pony started after the others, with me right behind, yelling and shooting. When we reached the rim of the valley I let him go. All four of them got the fright of their lives. When I reported the incident to the commander of a detachment of Chinese soldiers fifty miles farther on, he was furious because I hadn't killed at least one of the bandits. I told him, however, that I was a peaceable explorer and that it was his business to kill brigands, not mine.

But the real fun, as always came from the scientific work. The year before, when Shackelford found an extraordinary reptile skull at what we named the Flaming Cliffs, I sent it back to New York with him as one of our prize exhibits. A few weeks later I had an exciting cable from Professor Osborn telling us that we had discovered the ancestor of the Ceratopsians.

The Ceratopsians were a group of great horned dinosaurs, fossils of which suddenly appear in the upper strata of the Age of Reptiles in America. They were of large size, highly specialized, and evidently had gone through a long period of evolution. But where they came from, and when, was a mystery to science. In our little eight inch skull we had the answer. They came from Central Asia and we had found an early ancestor of the group. The species was named in my honor *Protoceratops andrewsi,* "Andrews's Before-the-Ceratopsian" dinosaur! It really should have carried Shackelford's name, for he was the original discoverer, but the leader of an expedition always gets the lion's share of credit.

Professor Osborn urged us to return to the Flaming Cliffs and learn more of the history of this important creature. In fact, he said, "If you don't go anywhere else the entire season you must get back there." Thus, the Flaming Cliffs was our main objective. We went through the desolation of a sun-parched desert from the Well of the Mountain Waters four hundred miles to the east. For a year there had been no rain. We followed the tracks of our own motor cars made ten months before. The scanty vegetation lay brown and shrivelled by the pitiless sun; white rims of alkali marked the beds of former ponds; the desert swam in maddening, dancing mirage that mirrored reedy lakes and cool, forested islets where we knew there was only sand. Mile after mile we traveled without seeing a living thing save scurrying spotted lizards and the long-tailed gazelles that do not need to drink. The way was marked by the skeletons of camels and the bones of sheep. The few Mongols with whom we talked before entering the desert told us that their friends had moved away from this area of desolation. Discouraged by the death of scores of ponies, sheep, and camels they had gone to the north in search of better feed.

Our caravan had been left near the Well of the Mountain Waters with instructions to follow us as fast as possible. Like all the camels of Eastern Mongolia, ours had suffered from lack of food and were woefully thin with soft, flapping humps. But old Merin thought at least some of them could hold out until they joined us in the Altai Mountains where, according to report, conditions were better. If they did not reach us the situation would be serious. Without gasoline, we would be almost as helpless as Robinson Crusoe on his desert isle. Nevertheless, I decided to take the chance.

Our arrival at the Flaming Cliffs was a great day for the Central Asiatic Expedition. We camped at three o'clock in the afternoon and almost at once the men scattered over the badlands. Before night everyone had discovered a dinosaur skull.

But the real thrill came the second day when George Olsen reported that he was sure he had found some fossil eggs. We joked him a good deal but, nevertheless, were curious enough to go down with him after luncheon. Then our indifference suddenly evaporated. It was certain they really *were* eggs. Three of them were exposed and evidently had broken out of the sandstone ledge beside which they lay. Other shell fragments were partially embedded in the rock and just under the shelf we could see the ends of two more eggs.

While the rest of us were on our hands and knees about the spot, Olsen scraped away the loose rock on the summit of the ledge. To our amazement he uncovered the skeleton of a small dinosaur lying four inches above the eggs. Almost certainly these were the first dinosaur eggs ever seen by modern human eyes.

In shape the specimens were elongated, much like a loaf of French bread, and were totally unlike the eggs of any known birds, turtles, or reptiles. Two of them, broken in half, showed the white bones of unhatched baby dinosaurs.

The preservation was beautiful. Some of the eggs had been crushed but the pebbled surface of the shells was as perfect as though they had been laid yesterday instead of eighty or ninety million years ago. Fine sand had filtered through the breaks and the interior of all the eggs was hard sandstone.

A later microscopical examination showed that the air canals in the shells are quite different in shape and arrangement from those of birds, turtles, or reptiles and proved that the white bone showing in the sandstone core of several of the specimens really are the skeletons of unhatched babies.

The small skeleton which Olsen had uncovered just above the eggs was a type of dinosaur completely new to science. It was only four feet long, although full grown, and toothless. Professor Osborn named it *Oviraptor* (the egg seizer) and he believed that it lived by feeding upon the eggs of other dinosaurs. Possibly, it was in the very act of digging up this nest when it was overcome by a sandstorm and buried upon the eggs it had come to steal.

A few days after the first discovery, five eggs were found in a cluster. Albert Johson also obtained a group of nine. Each member of the expedition became an enthusiastic egg hunter and everyone had success. Altogether twenty-five eggs were removed. Some of them were lying on the surface, exposed by erosion; others were enclosed in rock with only the ends showing and one nest in soft, disintegrated sandstone could be excavated with a camel's-hair brush.

The nights when we sat about a camp fire of tamarisk branches (a desert bush from the sand dunes near the tents), our talk was of dinosaurs and eggs. The deposit was unbelievably rich. Seventy-five skulls and skeletons were discovered, some of them absolutely perfect. Obviously the Flaming Cliffs were a region of great concentration for dinosaurs during the breeding season. Like living reptiles, dinosaurs scooped out shallow holes and laid their eggs in circles with the ends pointing inward; sometimes there were three tiers of eggs, one on top of the other. The lady dinosaur covered her eggs with a thin layer of sand and left them to be hatched by the sun's rays. She didn't sit on them like a hen. It was necessary for the covering sediment to be loose and porous in order to admit warmth and air, and it is possible that the exceedingly fine sand at this spot was particularly well adapted to act as an incubator.

I have been asked a thousand times since then if we expected to find dinosaur eggs when we went to the Gobi. I suppose that nothing in the world was further from our minds. As a matter of fact, we didn't even know that dinosaurs laid eggs. We supposed they did, for dinosaurs are reptiles and most reptiles lay eggs, but in the whole history of palaeontology no evidence of how dinosaurs produced their young ever had been found. We discovered the eggs purely by chance, in the examination of a deposit which we knew was rich in dinosaur remains.

Most of my friends seem greatly disappointed because our eggs are so small. They are only nine inches long and I've had a lot of explaining to do. Few people realize that there were big dinosaurs and little dinosaurs just as today there are pythons and tiny grass snakes. When the public sees a nine inch egg it is horribly disgusted. It demands something about the size of an office safe. It visualizes only the great Sauropod dinosaurs, *Diplodocus* or *Brontosaurus*, reptiles which could have looked into a second story window if there had been houses at that time. Those dinosaurs must have laid eggs, of course, and if they are ever found the public should be satisfied, for they ought to be a lot bigger than a football. But until that time the ones we have must do. After all, a nine foot dinosaur, mostly tail, could not be expected to do much better than a nine inch egg. That's a ratio of an inch of egg to a foot of dinosaur. Personally, I think it was a pretty good effort.

Every day of the second expedition was full of excitement and interest. The fossil localities, located in the first summer's work, all gave more important results than could have been expected. At Iren Dabassu the palaeontologists discovered a great quarry where bones of both flesh eating and herbivorous dinosaurs were piled one upon another in a heterogeneous mass. Geological and other evidence indicated that this spot had been a backwater or eddy at the edge of a great lake. When the dinosaurs died, their drifting bodies came to rest in the quiet bay. Then the flesh decomposed and the skeletons sank into the soft mud eventually to be fossilized.

My journal says of Iren Dabassu: "The number of dinosaurs that swarmed in this region during the Age of Reptiles, baffles the imagination. It must have been a nightmare country, filled with goblin-like creatures, stranger even than those born of delirium. Today, in place of this weird past world lie the silent, wind swept dunes of the Gobi Desert, parched and blistering under the summer's sun; in the winter an area of Arctic desolation. The alkali shores of a suncaked swamp mark a corner of the lake, the waters of which once lapped the edges of the ridge upon which we stood. As far as my eye could reach were hummocks of wind blown sand, crowned with thorny desert plants."

A dozen other spots yielded strange animals, some new to science; others known in Europe or America whence they had migrated from this Asiatic homeland millions of years before. But we could only skim the palaeontological cream. Most of the deposits would warrant years of exploration.

While the fossil hunters reaped their rich harvest, work just as important was carried on by the topographers who mapped great new areas; by the geologists, zoologists, and every other scientist on the expedition. But the making of a map, or the identification of a new geological horizon leaves a newspaper reporter cold and his editor completely frozen. So the dinosaurs and their eggs held the center of the stage, even as they did a hundred million years ago when the earth was young.

Chapter 22

The Valley of the Jewels

At the end of August we worked our way eastward and camped at the Valley of the Jewels. There Mac Young and I left the others to meet Professor and Mrs. Osborn in Peking. The Professor had traveled half around the world to visit the expedition in the field and I had selected a spot which was in the real Gobi but easily accessible from Kalgan. At the railway station in Peking we were met by Colonel Seth Williams of the U.S. Marine Guard. His first words were: "Yokohama and Tokyo have been destroyed by earthquake and fire. Hundreds of thousands of people have been killed. No one knows how many."

Only the bare news was known because the Japanese Navy, fearful that the United States might take advantage of the disaster to attack their stricken country, had jammed the air so full of conflicting radio waves that no messages could be deciphered.

As soon as word of the disaster was allowed to reach the outside world, American Naval vessels, loaded with doctors, nurses, medical supplies, and food, steamed at full speed from the Philippines to Japan. They were halted in the outer bay by Japanese warships and ordered to explain their presence. One of my friends was an officer on the American flagship. He was sent by our admiral to present his compliments to the Japanese commander-in-chief and inform him that the United States Navy wished to render aid. He found the Japanese vessels stripped for action with crews at quarters ready to give battle. It was many hours before Japanese suspicions could be allayed sufficiently for our men to begin their work of mercy.

I thought then, and have thought often since, what a revealing commentary it was on Japanese character. If they could believe that a friendly nation would take advantage of a great natural disaster to attack them, to kick them when they were down, it is just what they would do under similar circumstances. Otherwise it would not have entered their minds. Pearl Harbor has shown how true that was!

A short time later, when I visited Yokohama and Tokyo on the way back to America, I could hardly believe that such complete destruction of any city was possible. Not a single undamaged building stood in Yokohama. The city looked like the dump heap in a brickyard. Salvage had only begun. I stood on the ruins of the Grand Hotel and saw men take out the charred body of one of my intimate friends from a crushed bathtub. Martin, the hotel runner, was wandering about half crazed with grief, babbling, like a demented man, the names of old friends whom he had cared for there, and would never again welcome to his beloved hotel. I gazed sadly at the heap of ruin which had been Number Nine, knowing that somewhere in that pile of debris were the ashes of Mother Jesus, for, although many of the girls were saved, no trace of their mistress was ever found. And in Tokyo there was the great park into which thousands of people had rushed for safety only to be broiled alive by the blazing furnaces on either side.

Captain Robinson, one of the old time Pacific skippers, probably saw one of the most amazing spectacles any man has ever witnessed. He was standing on the bridge of the S.S. *Australia* which was to sail at noon. Orders had just been given to cast off the mooring lines when without warning the city collapsed like a house of cards before his eyes. A detail, he said, which stands out with photographic clearness was an automobile full of laughing Japanese racing down the wharf. Suddenly a great hole opened and it vanished from sight as though by magic.

The *Australia* herself was in grave danger. Captain Robinson told me that the lines from a capsized freighter had fouled his propeller and he dared not move the ship. Two Canadian sailors went over the stern and cut away the ropes just in time to avoid a great mass of blazing oil, acres in extent, which was moving swiftly toward the vessel on the surface of the water. Never, said he, would he forget that raging sheet of flame as it swept down upon thousands of men, women, and children struggling to swim away. He saw them caught and engulfed one by one in the red inferno.

The stories of escape by some of my friends were fantastic. The agent of the Great Northern Railway was just about to step into the back entrance of the Grand Hotel when the building suddenly fell away inward. A crack in the earth opened between his feet and closed again, leaving him unscathed. Because the quake happened just at noon many people had left their offices and were on the street, thus escaping death from the heavy tiled roofs which caused many casualties. The fire, of course, did most damage, especially in Tokyo, and a tidal wave sweeping in from the bay helped the destruction.

When Professor and Mrs. Osborn arrived at Peking, I had planned not to say a word about our discoveries until we had reached home, but before leaving the station platform the story was tumbling from my lips. Dinosaur eggs, titanotheres, coryphodons! The words poured out. Professor Osborn stopped, gasping, in the midst of a surging crowd of Chinese.

Both he and Mrs. Osborn were as thrilled as I had hoped they would be, not only by the expedition's extraordinary success but by our beautiful house and Peking's romantic surroundings. The morning after his arrival we sat for three hours over breakfast in the open sunlit courtyard, talking, while two crows perched on the roof tried to outdo us in a conversation of their own.

The next day, he and I, with Mac Young, left for Mongolia. The trip was perfect. Warm, windless days, brilliant sun and dancing mirage! The Gobi seemed to have put on its loveliest dress and best behavior in honor of our guest. It was a great moment for me when the man whose brilliant prediction had sent us into the field, reached our camp in the desert. The Professor, always full of sentiment, could hardly speak as he looked at the blue tents, the long line of camels, and the orderly row of cars.

"This," he said, "is the high spot in my scientific life."

The next days were the fulfillment of a dream to the Professor. Granger had discovered a splendid titanothere skull and left it in the ground partly exposed so that Professor Osborn might see, actually in position, one of the mammals he had prophesied would be found in Central Asia. He inspected all the important fossil localities in the Valley of the Jewels and at Iren Dabassu. A specimen in which he was particularly interested was a single tooth representing an archaic group of hoofed mammals called the Amblypoda. None of these great Ungulates had hitherto been known in Eurasia excepting *Coryphodon* of France and England. This single upper pre-molar tooth was the only specimen of the group the expedition had found in two years' search. We went over to the spot where it was discovered and later drove ten miles down the valley. When returning, Professor Osborn pointed to a low sandy exposure half a mile away and asked: "Have you prospected that knoll?"

"No, it is the only one in the basin that we haven't examined. It seemed too small to bother about."

"I can't tell why," said the Professor, "but I would like to have a look at it. Do you mind running over?"

When we stopped at the base of the hillock, I did not leave the car, but Professor Osborn and Granger walked out to examine the exposure. As he left, the Professor turned to me with a smile and said, "I'm going to find another *Coryphodon* tooth." Two minutes later he shouted, waving his arms. "I have it. Another tooth."

I could hardly believe my eyes and ears. Jumping out of the car I ran to the spot. The tooth I had discovered, five months before, was the third upper pre-molar of the right side. The one he just found was the third upper pre-molar of the left side and almost exactly the same size. Naturally, they could not have been from the same animal since the two were eight miles apart unless the beast just wandered about dropping his teeth all over the place. Of course, our scientific friends said we had planted the tooth for the Professor to find.

The last night we camped in an amphitheater surrounded by grassy hills. After dinner, Professor Osborn and I sat for an hour in front of my tent discussing the future of the expedition. It was obvious that our job could not be completed in the five years originally planned. Ten years at least would be necessary. That meant that I must return to America to raise another quarter of a million dollars. Moreover, it was highly desirable for some of our staff to go back to study and evaluate the work already done. We determined, therefore, to declare a recess in the field operations for a year and start anew in 1925.

Just as we were going to bed a dramatic incident gave the Professor a great thrill. Because our camp was in the bandit-infested grasslands of Inner Mongolia, I posted a sentry who was to be relieved every two hours. The candles were hardly out when he ran in to tell me that he heard horses. Galloping hoofs sounded plainly in the still night. I passed the word. When four men rode up, armed with rifles, they were quietly surrounded by all the men of the expedition. I ordered the visitors off their horses. Of course, they professed to be soldiers guarding the frontier but obviously they were bandits expecting easy pickings from a defenseless Chinese caravan. We collected their rifles and put them under guard. In the morning I released them but kept their guns, telling them that they could retrieve them at the military post at Chang Peh-hsien, where I would leave them.

When we reached there, the colonel told me the four men were well known bandits. He was very unhappy because we had not brought them in along with their rifles. A dozen soldiers mounted on fast ponies set out immediately, caught the men, and shot them before night.

In October, I sailed with the Osborns from Shanghai on a Dollar Line ship via the northern route. When we touched at Victoria, reporters from Seattle swarmed aboard. A representative of the *Post*

Intelligencer said, "I will give you fifteen hundred dollars for the exclusive use of the dinosaur egg photograph for a week." Another offered three thousand and a San Francisco paper upped it to five thousand dollars. I was aghast. From the foreign correspondents in Peking we knew that the dinosaur eggs had "caught hold" all over the world but expected nothing like that.

I told them, "I'll give you the news story but I'm not selling the photographs. After we reach New York I'll talk it over with the Museum authorities and see how best to give everyone an equal chance." Professor Osborn agreed that this was the thing to do but told me to use my own judgment.

At Seattle the mayor, the collector of the port, a vice-president of the Chicago, Milwaukee and St. Paul railroad, and a dozen other officials were on the dock to meet us. The transcontinental train was held half an hour while our trunks were being transferred. In the most painless way we found ourselves en route across the continent. These, some of the perquisites of having discovered a comic egg a hundred million years old! At every stop, on the way to New York, flashlights blazed and the photographers begged for "just one more."

It was very exciting and great fun although somewhat bewildering. New York was just the same only more so. But we made a great mistake about the dinosaur egg photographs. Wanting all the papers to have an equal chance, we passed out the photographs to all the reporters. Some of them were free lance photographers who promptly sold the pictures as "exclusive" to a dozen papers across the country. That experience was never duplicated. Always afterward, I turned the expedition photographs over to the *New York Times* "Wide World Service" who sold them on a syndicate basis. By this means they got fair and better distribution and the money from their sale went into the expedition funds.

The day after my arrival in New York, the late Mr. Adolph Ochs, publisher of the *Times*, asked me to luncheon in the beautiful dining room on the top of the Times Annex. Louis Ogden, editor-in-chief, was there with the brilliant John Finley and Arthur Sulzberger, present publisher of the *Times*.

My first visit was on a Saturday. The subject of radio came up and I happened to mention that I never had heard a broadcast. When I left America nearly three years before, radio was in its swaddling clothes. During that brief time, the baby had grown enormously and already was almost as commonplace as an automobile. Mr. Ochs was excited. "What! You never have heard a broadcast! Gentlemen, this will be a rare treat. He really has been out of the world, hasn't he? Right after luncheon we'll listen to the Yale-Harvard football game. Graham McNamee is reporting."

Mr. Ochs was as delighted as a child. When we went into one of the reception rooms he put me in front of the radio and the *Times* staff gathered round. Some years earlier an African explorer, Dr. Werner, had brought back a Congo pygmy for a psychological study of his reactions to the wonders of a great city. I knew then just how the pygmy must have felt. Mr. Ochs turned on the radio and suddenly I heard Graham McNamee's voice, vibrant with excitement, fill the room. I could hear the bands and cheering of the crowd. It seemed miraculous, incredible!

Then came the exciting reception which the public gave to our discoveries. John D. Rockefeller, Jr., and two of his children, with Professor James Breasted of Chicago came to my first lecture at the Museum. It was a wild night, with pouring rain and a full gale, but four thousand people tried to crowd into the Museum lecture hall which held only fourteen hundred. Motor cars extended in lines far up and down Central Park West across to Columbus Avenue and around to Eighty-first Street. I spoke for an hour while the overflow wandered through the Museum. Then the hall refilled and I gave the lecture over again.

Mr. Rockefeller was enormously impressed by this show of interest. A few days later he came to the Museum to see my specimens, and found the hall containing the dinosaur eggs packed almost like a subway train.

This time I had a splendid opportunity to show other parts of the Museum to him, none of which he had seen. As a result, some months later officials of the Rockefeller Foundation asked permission to make a survey and study of the Museum. They wrote an exhaustive report, which gave us high marks as an educational institution, and a million-dollar grant for our endowment fund.

Chapter 23

The Great Dinosaur Egg Auction

Dinosaur eggs! Dinosaur eggs! That was all I heard during eight months in America. There was no getting away from the phrase. Vainly did I try to tell of the other, vastly more important discoveries of the expedition. No one was interested. No one even listened. Eventually, I became philosophical about it. After all, the situation had its bright side.

I had returned in order to raise a quarter of a million dollars and the publicity was of great assistance. I might as well take advantage of it. How to do it? The people who had given me the first two hundred fifty thousand were pleased. Some of them, like Mr. Morgan, Mr. Baker, Mr. Rockefeller, and half a dozen others, doubled their original subscriptions. That helped but it was far from enough. The money was coming in too slowly for my peace of mind.

I was living with Professor Osborn at the time. One morning I said to him at breakfast: "I'm convinced the general public would help finance the expedition but they think small contributions aren't wanted. They believe this is only a rich man's show. If we could auction off one dinosaur egg as a contribution to the expedition's funds, it would be a grand publicity stunt. Every news story could explain that we've got to have money or quit work, that small contributions are more than welcome."

The Professor was just drinking a cup of coffee. He set it down suddenly. "Roy, it's a great idea. A ten strike. Let's do it."

Thus started the great dinosaur egg auction. A call went to the reporters to meet me at four o'clock in my office. Forty or more came. I told them frankly just why we were doing it and asked for help. "We'll sell the egg to the highest bidder. The proceeds will go to the finances of the expedition. Please stress in all your stories that it's up to the public whether we shall continue our explorations or not. Any contributions are welcome. I'll give you a report each day about what bids are received. They ought to make good stories."

The day after the announcement the *Illustrated London News* cabled an offer of two thousand dollars. The National Geographic Society upped it to three thousand. A museum in Australia bid thirty-five hundred. Yale University offered four thousand. The publicity was enormous and, true to their promise, the newspaper men included a plea for money in every story. Checks began to come in by every mail. Ten, twenty-five, fifty, a hundred dollars. Some were only a dollar; one was for ten thousand. By the time the auction ended, and Mr. Austin Colgate had purchased the egg for five thousand dollars as a gift to Colgate University, we had garnered fifty thousand dollars in public contributions.

I was very pleased but it proved to be a boomerang. Nothing else so disastrous ever happened to the expedition. Up to this time the Chinese and Mongols had taken us at face value. Now they thought we were making money out of our explorations. We had found about thirty eggs. If one was worth five thousand dollars, the whole lot must be valued at one hundred fifty thousand. They read about the other fossils—dinosaurs, titanotheres, *Baluchitherium*. Probably, those, too, were worth their weight in gold. Why should the Mongols and the Chinese let us have such priceless treasures for nothing? I had to combat this idea throughout all the remaining years of the expedition.

The assumption was natural, I suppose. They couldn't know that the five thousand dollars was a fictitious value engendered by publicity, or that any purely scientific or art object has a market value of just what it can be sold for to someone who has a special reason for desiring to possess it.

Meanwhile, the Lord had tempted me sorely. He had done so once before when oil and mining companies offered a hundred thousand dollars for the privilege of sending an expert with us. One of the biggest novelty manufacturers in the world came to my office with a letter of introduction from a prominent New York banker.

"I have a great plan," he said. "It will mean a lot of money for both of us. We'll make casts of the dinosaur egg, use them as paper weights, desk sets, etc., with your signature on the bottom. The original Easter egg! That's the idea. The first edition will be a million copies. I'll flood the world through my distributing agents. You can have a royalty. Most of them will be cheap. Sell for twenty-five cents. They'll go like hotcakes. You'll make a quarter of a million dollars or I'm a Chinaman. Here's a contract. You have your lawyer look it over and I'll give you twenty thousand dollars advance royalty the day it's signed. We ought to start production in two weeks."

It was a bit breath-taking. Attractive, perhaps, but even at first thought I didn't like it. That would be selling our birthright for a mess of pottage and no mistake! Of course, the money would be used to continue our explorations but everyone would think it went into my pocket. Moreover, the expedition would be stamped as a money-making venture in the eyes of the world. Science camouflaging business. Professor Osborn agreed with me but said, "Roy, this is your show. I'm going to keep my hands off. Do as you think best."

Just to satisfy myself, I asked twelve distinguished men in New York what they would do in my case. Eight of them said, "Let it alone. It's dynamite." The other four advised me to accept. I've always been glad the offer was refused.

Then came temptations of a different sort. Mr. Van Anda, managing editor of the *New York Times,* asked me to lunch with

him. "The *Times,*" he said "has had exclusive news service on all
the big exploring expeditions since Peary discovered the Pole. We
want yours. Whatever dispatches you send we'll give the place of
honor on the front page. We'll sell the English rights to the London
Times and to other papers throughout the world through our
syndicate. Not only will it give you enormous publicity, but you
can have all the proceeds from the syndication."

It was an attractive offer. The *Times* would handle the news in
its usual dignified way and give me a wonderful mouthpiece. But
there was a very big fly in the ointment. "You must remember," I
said, "that Peary, Shackleton, Amundsen, and all the others were
private individuals. The Central Asiatic Expedition is under the
auspices of the American Museum which is a public institution. We
depend for part of our support upon New York City. The news is
given out to all the papers equally. If our stories are released only
to the *Times,* every other paper will be furious. It would get the
Museum into too much trouble."

Mr. Van Anda didn't like it much. Professor Osborn was as
regretful as I, but there was nothing we could do. It would have
been a body blow to the Museum.

A week later, Professor Osborn gave a big dinner at the University
Club for prospective donors to the expedition. It was similar to the
one which had launched me on my money-raising campaign three
years earlier. Twenty or more of New York's greatest financiers
gathered in the Council Room to eat and drink and see movies of
our work in the Gobi. Among them was Frank Munsey, owner of
the *New York Herald.* I accepted an invitation to lunch with him
next day. I thought I knew what was on his mind. Sure enough,
the *Herald's* managing editor was at the table. Their proposition
was almost exactly that of the *Times.* Of course, I had to give them
the same answer. Munsey seemed very much annoyed but his face
cleared when I told him I had already refused the *Times.*

Then along came Mr. Hearst. He sent a representative to offer
me a quarter of a million dollars outright if I would give him the
exclusive newspaper, magazine, and movie rights of the expedition.

There was just one answer for that. No! My mind was made up before his representative had ceased speaking.

We used Dodge Brothers motor cars on our expedition. I purchased them in China and paid the regular price. At the end of the second year they were sold in Kalgan to Chinese importers of wool and furs from Mongolia. Strangely enough the old cars brought more than we paid. "After all," said the Chinese, "we know these cars can do the job because they've already been there. Perhaps new ones won't be as good."

Hardly had I reached New York when a representative of Dodge Brothers called upon me. They foresaw some priceless advertising. "I'll play ball," said I, "if you'll give us a new fleet of cars, made to our specifications. There are certain things we want changed. All in the body; none in the motor. I need eight new cars." They jumped at the suggestion like a trout taking a fly. Bayard Colgate and I went to Detroit to see Mr. Fred Haynes, then the president. He was a charming, white-haired man full of energy. His greeting was, "Now, gentlemen, Dodge Brothers employs twenty thousand men. You tell us what you want and we'll build it." That was the beginning of one of the most satisfactory business associations I ever have had. It continued even after Dodge Brothers was sold to Walter Chrysler and operated under the presidency of Mr. K. T. Keller, a man for whom I have great admiration. Their advertising was always dignified and thousands of cars were sold because of our endorsement. It saved the expedition about fifty thousand dollars.

Similarly, the Standard Oil Company gave us twenty thousand gallons of gasoline and five hundred gallons of oil, as well as supplying us with candles throughout the entire series of expeditions. I'm afraid the advertising they got in return didn't begin to repay their expenditure.

This trading of goods for advertising was entirely legitimate, I felt. We simply endorsed, publicly, products which had proved their worth on our first expedition. There were, however, critics who said we had sold out to Dodge Brothers and Socony. But it didn't bother me in the slightest. In a big public show one must expect criticism whatever one does.

From the time I first reached New York that winter of 1923-24 lecturing was the order of the day almost every day. It was the best possible way to spread the gospel of our expeditions and put me in touch with the people who might help finance them. But it engendered a bad habit. I began to talk incessantly and have never ceased. Now every time my wife and I walk down upper Fifth Avenue and see the electric sign "R.C.A." blazing in the sky, signifying Radio Corporation of America, she says, "That's you." My continual flow of words has been both a fault and a virtue. It bores my friends without doubt but it did make me a good salesman. I needed to be, for I had set my financial goal at three hundred thousand dollars.

One of my first lectures was at Smith, the famous girls' college in Massachusetts. It turned out to be an amusing experience. Arriving about five o'clock in the afternoon, there was just time to bathe and dress before dinner at six. I was horrified to find the guest apartment plastered with signs stating in no uncertain terms that the occupant must not smoke. That was bad news. I was smoking heavily then and just didn't see how I could speak without my usual cigarette. After dinner I excused myself, determined to do something about it. The guest suite was on the ground floor with an adjoining bathroom overlooking a path into the garden. Draping a blanket over the window, I stood on the "johnny," poked my head and shoulders outside, and lit a cigarette. I must have looked awfully funny. Just then a dozen girls strolled around the corner of the house. They came face to face with the distinguished lecturer of the evening breaking rules in what was far from a position of dignity. For a moment they gazed in amazement; then screamed with laughter.

"For heaven's sake," I begged, "don't give me away. If you tell on me I'm sunk after all those signs in the room."

They promised faithfully and I pulled my head in like a turtle retiring into its shell. When I went out on the platform half an hour later there were all my girls in the very front row. The introduction by the dean didn't help any for she stressed my scientific achievements to such an extent that one would have thought I must be as dignified as the archbishop of Canterbury. When I rose to speak the girls down front were grinning broadly. I made the mistake of looking at them and began to laugh in spite of myself. The faculty, sitting together, seemed puzzled and shocked. The rest of the audience giggled, too, and pretty soon the whole room was in an uproar although only the front row girls and I knew the joke. Remembering William Jennings Bryan's trick of twenty years ago, I pulled myself out of the hole by telling a funny story kept on tap for embarrassing moments and at last got into the lecture. The audience was grand. My girls in front applauded like a hired claque at every opportunity. It was, I suppose, the most successful public appearance of my career.

My lecture managers had arranged a trip across the continent. I spoke every evening and sometimes at schools and universities in the mornings and afternoons as well. Most of them, of course, were before women's clubs. Lecturers in America would starve if it weren't for the club women. Usually the "federation" arranges a course of lectures for the winter and spring. They pay big prices but the women make it their job to sell tickets. That winter Will Rogers, G. K. Chesterton, the English writer, and I were under the same management. Chesterton came first, next week Will and then I. Chesterton was an awful flop. He persisted in talking down to his audience as though they were a lot of first grade school children who wouldn't be able to understand a serious literary discussion. It didn't go. Poor Will Rogers got the backwash when he came along the next week. He used to leave notes for me. They were screamingly funny, the most priceless appraisals of the audiences and what they had said about Chesterton, written on a bit of newspaper or any scrap that happened to be handy. I would give much if I had them now. Of course, I saved them with his many letters, but the package was lost somewhere when I gave up my Peking house.

I had not, then, met Will Rogers and didn't, as a matter of fact, until eleven years later. He came out to Peking when I was en route to America and we passed each other in mid-Pacific, with an exchange of radiograms. I missed him again in California but we kept up a desultory correspondence for eleven years. We did not actually meet until 1935 when my present wife and I were on our wedding trip. From the Ambassador Hotel, I telephoned Will.

"Come right out to my house," he said. Then he proceeded to give a complicated set of directions which were mostly "Don't turn in there 'cause that's not my gate." Finally, "Oh, hell, I'll meet you at the studio for lunch. Then we'll be sure not to miss each other again." When Will got a look at my wife he said, "Where did you get her? You didn't dig her up out in the Gobi! If you did, I'm going."

The lunch was with Will and John Boles and Conrad Nagle at the studio restaurant. Then we went out to Will's house. There was a lot of construction going on. Will said, "Mrs. Rogers is away. We never can agree on what to build. So when she goes, I have a fling. When I'm away she does what she wants to the house. She only left a week ago. I had five carpenters hid out in the bushes. As she left the driveway I yelled, 'Come on, boys, get to work.'"

We went over to the Uplifter's Club where Will played polo that afternoon and joined Irvin Cobb, an old friend of mine. Will and Irv together were priceless. I never saw Will Rogers again. He told us then of his intended trip with Wiley Post which ended in tragedy on the stark Alaskan shores only a few months later. The country could ill afford to lose him. He formed a link between the government and the people with his sanity and homely philosophy that is sadly needed in these hectic days. He never seemed to lose his perspective or kindliness no matter what the situation.

On that lecture trip in 1924 John Barrymore met me at the station in Los Angeles, much to the delight of the news photographers. John was at the height of his glamor and fame. He had just finished a season in *Hamlet* which made even the most hard-boiled critics rave. Under his wing I saw the best of the fantastic movie community.

John had a collection of weird and strange objects from all over the world. Naturally, he was wild to possess a dinosaur egg. I couldn't let him have one but did give him a dozen bits of shell about the size of my thumb nail. He kept them in a glass-top box along with a letter of authenticity from me. Some ten years later a syndicated story appeared about the enormous dinosaur egg I had given Barrymore. The only one in private captivity. John was very much in eclipse at the time and I guessed that his publicity man in casting about for stories with which to get the actor's name back in the papers had discovered the bits of shell or the letter. Of course it was nothing for a press agent to make them into a whole egg bigger than one ever dreamed of by a dinosaur. After Barrymore's death a gentleman in California wrote me to the effect that he had bought my letter to John at an auction of his effects but wasn't able to find the egg to which it referred. Could I enlighten him? "Like everything else that was sold," he wrote, "it was said to have come from his bedroom!"

By the time that lecture trip was ended, I felt and looked like a sucked orange. During four months I had appeared before a hundred and twenty-five audiences. But it had produced nearly a hundred thousand dollars for the expedition. That wasn't mere accident. I worked it all out in advance. In every city where I appeared my secretary had prepared a carefully edited list of the people who had the means to assist in financing the expedition. Usually someone of the prominent citizens wanted to give me a dinner. I indicated, without too much subtlety, that I'd accept if so and so were present. Almost always so and so was there. Then I used the propaganda that experience had taught me would be most effective. Rich women of middle age were my best clients. I was completely shameless, I'll admit, but it was a worthy cause. My job was to raise three hundred thousand dollars for the Central Asiatic Expedition, and I never forgot it. Believe me, it wasn't all fun. By the time I returned to New York I was so exhausted, nervously and physically, that the one thing in the world I wanted was the peace and quiet of the Orient, where I could play polo and forget that money existed. When the last check came in and I knew that I had considerably more than a quarter of a million dollars in my trousers' pocket, I sailed for China.

Chapter 24

Desert Dune Dwellers

Just before sailing from San Francisco in 1924, the American minister to China cabled that his daughter and niece were to be on the S.S. *President Cleveland* and would I chaperon them to Peking? They were nineteen and twenty-three, respectively, pretty, vivacious, and intelligent. It took me about three seconds to introduce myself and we did a turn about the deck with newspaper men hot on our trail. Jim Corbett, who was a friend of mine, had come to see me off. That was manna from heaven for the photographers. Ex-world's champion prize fighter, daughter and niece of American minister, Gobi explorer, all in one picture! Then I discovered another friend, Frank Buck. He was sailing with us on the *Cleveland* to bring back a cargo of wild animals from Singapore. Before the vessel was past the Golden Gate, our foursome was well established. That trip across the Pacific was fun, and I delivered my charges to the American minister in due course.

Back in Peking the days were crowded with interest. An international commission had arrived to discuss the Chinese Customs which paid the Boxer indemnity. Silas Strawn, a prominent lawyer of Chicago, headed the American delegation. We became great friends. He was far and away the most brilliant man at the conference.

The late Lord Willingdon arrived in Peking at the same time. He had recently retired as governor of Madras, India, and came to China to determine what the British would do with their share of the Boxer indemnity.

After a time, I went off to Urga, with Gordon Vereker, brother-in-law of Lord Gort, to make diplomatic arrangements for our next expedition in the spring of 1925. Urga was an awful place. The Russians had taken over with only a semblance of maintaining Mongolian autonomy. Behind each cabinet minister stood a Russian "adviser" who told the poor Mongol what he could and could not do. Getting into Urga was an ordeal of questioning and secret police. Getting out of it was infinitely worse. It took weeks to satisfy them that we should be allowed to proceed with the Central Asiatic Expedition. They finally consented only when I agreed to give them half our finds and take two "representatives" with us to report on our activities. It was pretty bad compared with our procedure of only a few years before when the Mongols had been free dwellers in the open plains. Some of my old friends, the premier and one or two others, in the privacy of their yurts, opened their hearts to me. For these proud Mongols to be reduced to virtual slavery was pitiful. But what could they do?

A very bright spot in that visit to Urga, however, was meeting General P. K. Kozlov. As a young man he had been with Prjevalsky, greatest of all the Russian Asiatic explorers. Trained in this hard school, Kozlov had enrolled his own name in the "immortals" of Tibet and Mongolia. Kozlov was nearly sixty-five at the time, but with all the enthusiasm of a boy he was preparing an expedition to re-excavate Kara Khoto, an ancient city buried in the sands of south central Gobi. The Russians had refused to let him proceed, but Kozlov turned the delay to advantage by discovering a dozen remarkable tombs of the Tang Dynasty in the forests north of Urga. With him I visited the excavations. His charming wife, thirty years his junior, was living at the tombs in a log hut. She spoke perfect English but Kozlov knew not a word.

With her to translate, when my bad Russian failed, we talked of early days in Central Asia, of his plans and mine. I wondered where the money to carry on his work came from. Madame Kozlov hinted that he had sold in Amsterdam, secretly, a wonderful diamond the czar had given him. That was, she thought, the reason the Bolsheviks would not let them go to the "Black City" in the Gobi sands. I saw Kozlov only once after that, in Urga. He is dead these dozen years. The Russian press gave him his just due as a great explorer even though he remained always loyal to the czar.

The 1925 expedition was the largest and most ambitious of the series. In fact, it was too large, I discovered, for completely effective work. Fifty men all told, eight motors, and one hundred and fifty camels. It took so much food, so much gear, and so many cars to handle the big party that our mobility was somewhat reduced. Nevertheless, it proved to be one of the most successful seasons.

I increased the number of sciences represented to seven. We had become greatly interested in the climates of Mongolia during successive geological periods and wished to check our deductions with a study of fossil plants and insects as well as from geology and palaeontology. Osborn's original thesis that Central Asia was a great center of origin and distribution of northern animal life had been demonstrated pretty clearly. But other important facts emerged. The Central Asian plateau was the oldest continuously dry land in the world. For a hundred and fifty million years it had been rising while Europe and America in part had successively risen and sunk below the sea. But Central Asia furnished an unbroken record of animal life such as existed nowhere else in the world.

The Mongolian plateau never had experienced an invasion of ice as did Europe and America. There were climatic cycles, wet and dry, wet and dry; also cycles within cycles. These fascinating problems made the year's work important far beyond the limits of pure palaeontology. The accuracy of our maps, too, never had been approached in former explorations.

That season was similar to the first two. New country explored, new areas mapped, new fossil deposits opened. Greater difficulties from the changed political conditions required tact and force at times. Bandits were a nuisance, but in two or three "incidents" we lost no men.

At the Flaming Cliffs more eggs were found, bigger and better eggs. Also there came a fortnight of intense excitement when traces of a previously unknown human culture were discovered in the basin less than half a mile from camp. It was an area of shifting sand blown into dunes against the stems of twisted tamarisk trees. Sculptured red bluffs marked the entrance to shallow valleys floored with sandstone where the wind had swept the loose sediment away.

On the clean hard surface of the rock pointed cores, tiny rounded scrapers, delicately worked drills and arrowheads of red jasper, slate, chalcedony, and churt were scattered like newly fallen snow. Among them were bits of dinosaur egg shell, drilled with neat round holes—evidently used in necklaces by primitive peoples. Also there were pieces of crude pottery. It was a strange jumble of conflicting specimens, some indicating an Old Stone Age culture—others dating the time of their deposit there as much later. After two weeks of intensive work certain pretty definite facts emerged. The site had been used continuously by human beings for thousands of years. The oldest culture was Palaeolithic, then came a transition stage which gradually developed into the Neolithic.

It is believed that most of the primitive races whose remains have been discovered in Europe came from Asia, and that wave after wave arrived from the East, each one driving out or annihilating the people they found in possession of the region. Many of these races left stone tools or weapons, highly characteristic of their particular cultures. The problem confronting us was to determine where our people fitted into the mosaic of primitive European humans.

As a matter of fact, they represented a hitherto entirely unknown race, with a culture distinctly their own, most closely allied to the Azilians of France and Spain. We named them the Dune Dwellers. They roamed over all Mongolia ten to twenty thousand years ago. Dressed in skins, probably living under crude shelters of hides or bushes, they hunted, fought, and loved much as do the primitive savages of Australia and Tasmania today.

There was another great discovery at the Flaming Cliffs that summer. In our first year's collection, Granger had labeled a tiny skull "unidentified reptile." Eventually, it was freed from rock and Dr. W. D. Matthew wrote excitedly: "It is one of the earliest known *mammals.*"

Bryan and all his cohorts to the contrary, we know that out of cold-blooded, egg-laying reptiles evolved the warm-blooded mammals which give birth to their young alive and nourish them with milk. In a hundred years of scientific research only a single skull of a mammal from the Age of Reptiles ever had been discovered. That was *Tritylodon* of South Africa. The British Museum considered it one of the world's greatest palaeontological treasures. At the Flaming Cliffs we found seven skulls and parts of skeletons of these Mesozoic mammals. They were tiny creatures not larger than a rat and crawled about in the midst of dinosaurs at the close of the Age of Reptiles a hundred million years ago. After the dinosaur eggs have been forgotten these little skulls will be remembered by scientists as the crowning single discovery of our palaeontological research in Asia.

We left the Flaming Cliffs with regret. They had given us more than we dared to hope from the entire Gobi—dinosaur eggs, a hundred skulls and skeletons of unknown dinosaurs, seven Mesozoic mammals, and the new Dune Dweller human culture. As my car climbed the steep slope to the eastern rim I stopped for a last look into the vast basin, studded with giant buttes like strange prehistoric beasts carved from sandstone. There were medieval castles with spires and turrets, colossal gateways, walls and ramparts; caverns that ran deep into the living rock, and a labyrinth of ravines and gullies. I would never see them again. "Never" is a long word but I knew that for the last time my caravan had fought its way across the desolate reaches of the Gobi to this treasure vault of world history.

We were driven out of our final camp of the season by a plague of snakes. Pit vipers are the only snakes in the desert and they are extremely poisonous. Our tents were pitched on a promontory with steep rocky sides. During the night, the temperature dropped to freezing and the vipers came up to get warm. Norman Lovell was lying in bed when he saw a wriggling form cross the triangular patch of moonlight in his tent door. He was about to get up and kill the snake when he decided to have a look before putting his bare feet on the ground. About each of the legs of his camp cot a viper was coiled. Reaching for a collector's pick ax, he disposed of the two snakes which had hoped to share his bed. Then began a still hunt for the first arrival. He was hardly out of the sleeping bag when an enormous serpent crawled from under a gasoline box near the head of his cot.

Lovell was having rather a lively evening of it but he was not alone. Morris killed five vipers in his tent and Wang found a snake coiled in his shoe. Then he picked up his soft cap from the ground and a viper fell out of that. Dr. Loucks actually put his hand on one under a pile of shotgun cases. Forty-seven snakes were killed that night. Fortunately, the cold made them sluggish and they were slow to strike. But it got on our nerves and everyone became a bit jumpy. The Chinese and Mongols deserted their tents, sleeping in the cars and on camel boxes. The rest of us never moved after dark without a flashlight and pick ax. When I walked out of the tent one evening, I stepped on something soft and round. My yell brought out the whole camp but it was only a coil of rope. A few moments later, Walter Granger made a vicious lunge with his pick shouting, "I got you that time!" But Walter had merely sliced a pipe cleaner!

The new camp proved to be as rich in fossils as in reptiles but at last the snakes won. Moreover, flurries of snow warned us to be on our way southward. On September 12, the cars roared down the slope to the basin floor, leaving Viper Camp to the snakes and vultures.

Chapter 25

Motoring through a War

Through the gates in Peking's Tartar Walls came many of the world's great and near-great. One of them, the late Lord Northcliffe, England's leading newspaper publisher, paid us a visit. I remember him chiefly as the rudest man I ever met. Northcliffe was staying at the British Legation. The minister, a great friend of mine, suggested that I give him a dinner as he wanted to see some of our specimens. That night I felt rather pleased for the house looked lovely and my guests were the most interesting of Peking's cosmopolitan residents. As I remember, there were nine nationalities represented in the twenty-two people at the table. The cook had done himself proud and with the roast wild duck we had some special muscatel from one of the Belgian missions in Mongolia. The wine was made in Algiers, taken to Rome to be blessed by the pope, and sent to the Far Eastern missions for use in religious celebrations. I got a cask of it by trading many cases of beer to the priests at Hei-ma-ho, north of Kalgan. I don't know how long it had been aged, but it was something to dream about.

Everyone felt very comfortable, and as we were having coffee and port after dinner the general commanding the British Air Force in China started to tell a story. Apparently it bored Lord Northcliffe, for suddenly he said, "That's quite enough from you," and abruptly turned his back on the narrator. The general's face flooded with color and for a moment I thought we were to have a scene but he controlled himself, signaled to me, and I walked with him to the door. "Sorry, old chap, but I've got to go. You heard what the blighter said. If you don't mind, I'll just slip off."

I was outraged that Northcliffe should have been so impolite to one of my guests and particularly to the ranking British general. He didn't improve matters any when a little later I was showing him our specimens in the laboratory with the British minister. Suddenly Northcliffe turned to me.

"Dr. Andrews, I'll give you ten thousand pounds for the exclusive rights to your stories for my papers when you next go to Mongolia."

"But I can't do that," said I. "There would be a devil of a row in America if the news about our work had to come there from London. Moreover, I was offered fifty thousand pounds in the United States and turned it down because I can't give exclusive rights to any newspaper."

Northcliffe didn't press the point but the rest of the evening he was as petulant as a small boy who'd lost a stick of candy. He only grunted when he said good night to me. His whole visit to Peking was a succession of unfortunate incidents and everyone was glad to see the last of him.

Noel Coward delighted Peking. He came with a young British peer, whose name I have forgotten, and both threw themselves wholeheartedly into the life of the foreign community. My police dog, Wolf, was most choosy in his friends. The first day Noel came to my house he and Wolf had a little private conversation in which they established a complete understanding. Ever after Wolf went into transports of delight when Noel entered the compound gate, and before long they would be rolling on the floor like children. Noel was gay, friendly, and obviously crazy about Peking, which was his best passport to our affections.

Douglas Fairbanks, arriving a few weeks after Noel, was another man who fitted into our cosmopolitan community. Douglas and I had become friends in Hollywood. He said to me once, "You are the only man in the world whom I envy. You do in real life the sort of things I have to simulate in pictures. I'd give my soul to be an explorer." He'd have made a good one.

His trip to China initiated a period of frenzied traveling for Douglas. He was wild to kill a long-haired Manchurian tiger and I told him I'd arrange a shoot. He couldn't make a definite date, however; he'd have to come when he could get away. As bad luck would have it he cabled me, "Coming for the tiger," just as I left for America. Our ships passed in mid-Pacific and we had a wireless exchange with Fairbanksian abandon; all about tigers, what to do and whom to see in China. No one of prominence who came to the Orient in my time was so genuinely popular as was Douglas Fairbanks.

I stopped in New York only long enough to say hello at the Museum and then hopped over to England on the S.S. *Aquitania* to do a lecture at the Royal Geographical Society in London.

Back in New York I had a new sort of adventure and one which proved to be very amusing and instructive. I had my first operation. On New Year's Day, 1925, in the annual point-to-point hunt at Tientsin my pony came down at a jump, and I got a dislocated right collar bone. For six weeks I lived in a cast, very much disgusted, but the repair process did not go just according to plans. All summer in Mongolia there was a lot of pain and it seemed best to do something about it. Dr. Fordyce St. John, an old friend, and incidentally one of the best surgeons in America, told me a prong of bone was growing from the clavicle toward the brachial plexus and that in about a month I'd suddenly drop off unless it were removed. It upset the lecture schedule completely and my managers were wild, but even they had to admit that a dead lecturer would not benefit their exchequer. So I went into the old Presbyterian Hospital on Park Avenue and Johnny St. John did a job of work on me. It was like a new exploration. I'd never had an operation or an anesthetic and knew nothing about the mysterious workings of a hospital where people really are sick. I determined to enjoy the experience. The Presbyterian, at that time, was a homey, informal sort of place, very different from the great medical center of today with which it merged.

"I'm going to be weak but not sick," I said to Johnny, "so I want to have some fun. A pretty nurse will help a lot. I'd like to pick her myself." He agreed and put me in the way of doing so at once. It turned out to be a very serious operation and for several days I wasn't interested in anything. Then I began to enjoy myself. Friends sent in all sorts of delicacies; Johnny brought me fresh eggs from his farm and I lived the life of Riley. My room became a gathering place for the internes and nurses off duty. I was beautifully looked after, had excellent food, and continual entertainment! What more could a rising young explorer ask?

After about ten days, George Palmer Putnam appeared. I had spent a weekend at his house just before going into the hospital and in a long morning's walk proved conclusively that I did not have time to do a book he wanted on the first three years of the Central Asiatic Expeditions. I remember he agreed it *was* impossible but just as we were on the doorstep he said, "Now that's all settled. When are you going to start the book?" There is nothing one can do about a man like that. "You," said I, "can go to hell." And went in the house.

The day George came to the hospital I was holding a rather larger court than usual. He waited until the others had gone.

"Now, Roy," said he, "instead of wasting your time sitting here like a Turk in his harem, you can write that book. I'll send you up an assistant tomorrow. You can dictate a lot of it. Here's a check for a thousand dollars advance royalty. Will you do it?"

It is always hard for me to refuse anything to George for I am very fond of him. Moreover, what he said made sense. Next morning Mrs. Charlotte Barbour arrived and stayed until I left the hospital. Gone were my days of play. She and George bullied me unmercifully and kept my nose to the grindstone. At last Professor Osborn took me to his apartment for the final convalescing. Charlotte came every day. By putting together a lot of articles I had written for *Asia* magazine and dictating other chapters, the book was completed in a month. The title is *On the Trail of Ancient Man*. It isn't a very good book but not bad either, as my books go.

Somewhat wooden in spots. Still, it does tell the story of those first three years well enough.

I began to fill lecture engagements while still too weak, and at the first one in the Brooklyn Academy of Music I fainted on the stage. I came to consciousness in the dressing room and was horribly embarrassed. After a few minutes of rest they pushed me out in a wheelchair and I finished the performance sitting down. There was, too, a very satisfying lecture at Carnegie Hall. The place holds twenty-eight hundred people and it was packed to the ceiling half an hour before the lecture began. So many were unable to obtain seats that my managers decided to repeat the lecture next day as a matinee. With only an announcement from the platform the hall was filled. I was pleased—and you may be sure the managers were!

Until early spring, I did nothing but talk from one end of the country to the other. It netted fifty thousand dollars for the expedition. I sat at so many dinners and met so many interesting and important people that they formed only a blur of places and personalities. Ending up on a Pacific liner, I slept most of the way across to Japan.

Bad news awaited me in Kobe. Feng Yu-hsiang, the so-called "Christian general," was having rather a lively war with Chang Tso-lin. It had started the previous October down near Shanghai and had spread like a flame to the north. The severest fighting took place during December and January near Tientsin. By April first, when I arrived, Chang had occupied Tientsin but Feng had fallen back upon Peking.

For weeks there had been no railway traffic between the two cities. The foreign powers protested. The Chinese told them to go jump in the lake, or words to that effect. That was all wrong, for the 1900 treaty provided that communication from Peking to the sea should be kept open. Still the fact remained that the only way to reach the capital was by the motor road.

McKenzie Young drove down to meet me. On the return trip we discovered that the road had been mined in thirteen places. It was a rather jumpy business. Leaving Tientsin we passed out of Chang's

lines through a wide "no man's land" and into the rear guard of
Feng's army. No active fighting was going on at the time but it
was a decidedly unsafe road to travel. Soldier stragglers roamed
all along the way, robbing whenever they had the opportunity. But
we were lucky and reached Peking without serious difficulty.

Two days later I was dining with the American minister when
firing began just outside the city and we all adjourned to the roof
of the Peking Hotel. Machine guns showed in a steady stream of
light along the southern horizon, punctuated by the wide flashes
of heavy guns. The American military attaché told us that Feng
had begun a new offensive and might even push Chang's army
back to Tientsin. But one of Feng's generals was bought off by the
opposing side and the advance became a retreat.

I had to get through to Tientsin to bring up Dr. W. D. Matthew and
made a try for it the next day with three members of the expedition,
Shackelford, Hill, and Beckwith. We thought that a large American
flag on the car would protect us as it had in former years.

The gates of Peking were heavily guarded but the soldiers let
us pass. Carts were already coming into the city with grain, camp
gear, and soldiers. Cavalry streamed by and then thousands upon
thousands of infantry. They were retiring in good order and seemed
most cheerful. An officer told me that Chang Tso-lin's troops had
taken Tungchow, fourteen miles from Peking, and were looting
the city but that no fighting was taking place.

We drove on slowly and eventually passed beyond the rear of
the retreating army. For three or four miles the countryside was
deserted, houses closed and all as quiet as the grave. We were five
or six hundred yards from the ancient marble bridge at Tungchow
when there came the sharp crack of a rifle and a bullet struck beside
the front wheel. A second later a mass of soldiers appeared on the
road and bullets began spattering around us like hailstones. They
had opened fire with a machine gun but it was aimed too low
and the lead was kicking the dust just in front of us. The soldiers
could see the American flag plainly enough but that made not the
slightest difference.

Fortunately at this particular spot the road was wide enough for the car to be turned and I swung it about in record time. The bullets were now buzzing like a swarm of bees just above our heads. Forty yards down the road a sharp curve took us out of sight of the machine gun. The other men crouched in the bottom of the car. Since I was driving I could see all the fun. It was a pretty rough road but the speedometer showed fifty miles an hour as we went back.

The ride became exciting. All the houses which had seemed so peaceful actually were occupied by the advance guard of Fengtien soldiers. They had let us pass because of the American flag but when they heard the firing in our rear and saw us returning at such a mad speed, they evidently thought that we were anybody's game. Each and every one decided to take a shot at us.

For three miles we ran the gauntlet of firing from both sides of the road. I would see a soldier standing with his rifle at the ready, waiting until we came opposite. Then "bang" he'd let us have it. Sometimes they fired in squads, sometimes singly. But they were bad shots. Most of them aimed directly at the car, when they aimed at all, and the bullets struck just behind us. Every now and then one would zip in close to my head but no one was hit. I really had the best of it because the others could not see what was going on and driving the car kept me busy. I expected every moment that one of the tires would be hit. A blowout at that speed might have turned us over.

Before long we could see the rearguard of the retreating army and the sniping ceased. Still our troubles were far from ended. The first soldiers, three of them, asked for a ride. I thought that they might be a protection and let them stand on the running board of the car. Suddenly one of them saw an officer. Without a word he stepped backward off the car, and rolled on the ground with his right hand under the rear wheel. As I put on the brakes it ground his hand and arm into the hard gravel road. I have never seen such a sight. His hand was simply shredded. I put on a tourniquet to stop the bleeding but he was only anxious for us to go on.

We went but soon had to slow up because of the masses of infantry. Soldiers began to jump on the car. Both running boards were jammed solid, others hung on the rear, two sat astraddle the hood. I couldn't see to drive. The car could barely crawl in low gear. Vainly I protested. These men spoke the Shantung dialect which is difficult to understand. Then came an accident that set off the fireworks. One of the men on the hood fell off. A wheel ran over his leg and the heavy load plus the gravel mangled it badly.

The Chinese have a great tendency to talk themselves into a rage. They yanked us all out of the car, shouted, and gesticulated. Finally hysterical with anger, they lined us up against the car and cocked their rifles. Things looked pretty bad. Just then an officer appeared. Fortunately he could speak Mandarin Chinese and I explained what had happened. He said, "I am a staff officer but I can't control these men. You must get off the road at once or you'll be killed. Drive down the bank there into the fields. I'll stay here until you are out of rifle shot."

It was a difficult job to navigate over the plowed ground, but somehow we got to the gate of Peking and into the city. The experience affected each of us differently. I had been so busy driving that there was no time to be scared, or at least not to give in to the feeling. I had got the other fellows into the jam and had to get them out. But once back in Peking I felt awfully weak and sick. One of the other men who was staying with me had been perfectly cool throughout the entire performance and afterward. At two o'clock the next morning he went into violent hysterics. I had a beautiful time getting him back to normal.

Chapter 26

Politics and Palaeontology

The years 1926–27 marked a turning point in the history of modern China. It was vitally important, too, in the fortunes of the Central Asiatic Expedition. Anti-foreignism burst into flame out of the still smoldering fires of the Boxer Rebellion of 1900. Foreign concessions, extra-territoriality, and the domination of the white man had been endured by the Chinese only because they could not resist. The foreigner was a "hair shirt" to their sensitive pride.

Chiang Kai-shek, in command of his Yangtze Valley army, was energetically making plans to attack the North under Chang Tso-lin with the object of bringing the entire country under the control of his party, which had been getting assistance from the Russian, Borodin, and Russian generals.

The Chinese eventually moved on Shanghai with the avowed intention of taking that city, ousting the foreigners, and claiming the International Settlement for the Chinese. Flushed by their unexpected success in regaining the Hankow Concession, the Chinese decided that the Powers would not use force and they could drive all foreigners out of the country. They demanded the return of the Tientsin concessions and of the Legation Quarter in Peking. Propagandists, directed by Borodin, were active in all parts of China and particularly in the Northern armies.

But the Foreign Powers had been driven too far by the advance upon Shanghai, and war vessels and troops began to arrive from all quarters of the globe. The British sent battleships and a large force. One of the first regiments upon the scene were the Coldstream Guards with Lord Gort, later to become famous at Dunkirk, as colonel. It is generally admitted that their prompt action prevented a most horrible massacre of foreigners.

An indication of what would have happened all over China was given at Nanking. A Chinese army killed half a dozen foreigners, attacked several consulates and looted the city. The resident foreigners gathered on a hill belonging to the Standard Oil Company where they were besieged. From the roof of the house a gallant American Marine signaled the warships in the river while bullets spattered all about him.

The officer commanding the American destroyer had been ordered not to fire on the Chinese. But this was a crisis. With the remark "I'll either get a medal or a court martial, but here she goes," he laid down a box barrage about the hill. At the first high explosive shell the Chinese ran pell mell. A landing party rescued the besieged foreigners.

The river at Shanghai swarmed with ships. It was said that never before in history had war vessels representing so many nationalities anchored together in one harbor. Barbed wire entanglements were erected and the International Settlement put under martial law. One or two clashes took place with considerable loss to the Chinese but no determined attack was launched against the Concession. All foreign legations ordered their nationals from the interior of China. Reports were continually arriving of murders and outrages committed on foreigners in various parts of the country. It was a repetition of the events leading up to the Boxer Rebellion of 1900 only on a wider scale.

For the first time I saw something like a panic in Peking. Even the year before when the gates were closed and sandbagged and Chang Tso-lin's wild Manchu hordes were looting and burning the countryside, few foreigners were even nervous. But the Nanking outrage made us realize that wholesale slaughter had been averted only by the arrival of foreign troops. The Southerners were pushing slowly northward and all legations advised their nationals to leave Peking. Women and children went to Dairen, Japan, or Manila. We had a carefully thought out plan for the protection of the expedition's headquarters. With machine guns posted on the roofs we would be able to present a pretty strong defense against looters or even well armed soldiery. All the staff offered to remain with me and protect the house if necessary. Notices were sent to foreign residents in Peking where to assemble in the event of extreme danger. Green lights and guns were the designated signals. I told the American Legation that we were not going to leave our house and valuable equipment; that was certain.

Suddenly the situation was completely changed by Chang Tso-lin's dramatic raid on the Dal Bank and the Russian Military Attaché's office next to the Soviet Embassy. The raid took place at eleven o'clock in the morning with the unofficial permission of the diplomatic body. I happened to be at the National City Bank on the opposite side of the street and witnessed the entire proceeding. It was most spectacular and totally unexpected. Even in his wildest dreams Chang Tso-lin could not have believed that the results would be so important. The Russians had depended upon the diplomatic immunity of the embassy and used it as a central clearing house from which operations were directed all over the world. Raids which subsequently took place in London, Paris, and the Argentine were made upon information obtained at the Soviet Embassy in Peking. Chang Tso-lin then set to work systematically to rid North China of propagandists. Those Chinese who were caught in the Russian Embassy raid were slowly strangled in front of the Old Marshal himself. Hardly a day passed that one or more persons were not executed at the public ground opposite the Temple of Heaven.

The prospect for continuing our explorations in Mongolia could not have been blacker. We proceeded to liquidate certain effects of the expedition, put others in a place of safety, and reduce current expenses to the minimum. All the staff, except McKenzie Young, I sent back to America. He remained with me to watch events and prepare for an expedition in 1928 if it became possible. In Peking, conditions settled down to normal with amazing rapidity after the Soviet Embassy raid. War, of course, was still going on in Central China—but then, there was always war! So long as it didn't come knocking at our front door, it was only of academic interest. The flare-up at Shanghai was a Heaven-sent excuse, however, for congressmen from the United States to "come out and see for themselves" what they ought to do about it. They descended on us in droves.

During the Shanghai affair the Third Brigade of U.S. Marines came out with Brigadier General Smedley Butler in command. Later it was transferred to Tientsin. "Smed" was the "stormy petrel" of the Marine Corps but a magnificent officer with a distinguished record and a man who believed that the function of the Marines was as much to keep peace as to make war. I often told him that he was a hundred horse power motor in a Ford chassis. His tremendous nervous energy simply burned up his frail body. It gave him devastating headaches and nervous indigestion. But with all he carried on and did a job in China of which the Marine Corps and all Americans can well be proud.

Tientsin was threatened by a fire which started near the tanks of the Standard Oil Company where enormous quantities of gasoline and oil were stored. A strong wind was blowing directly toward the city. If the tanks exploded, nothing on earth could save Tientsin. General Butler never waited for a request. He called out the Marines, took personal charge, and by their initiative and bravery saved the city and thousands of Chinese homes and people.

Smedley Butler and I became great friends. He used to motor up to Peking to spend almost every weekend with me and I never could hear enough of his stories of action in the Great War and campaigns over the world. During the Boxer Rebellion of 1900 he had been with the U.S. Marines as a young lieutenant and was wounded by a sniper when they attacked the southeast gate of the city. One day Smedley showed me the exact spot on the wall where the Chinese stood and the tree beside which he fell when the bullet struck him. Nothing, he said, had changed since that day twenty-six years before.

The motor road between Peking and Tientsin was in a deplorable condition. One Saturday the general found that a small bridge at Yang-tsun, ten miles out of Tientsin, was gone. He drove back to the barracks and, as a practice job, ordered his engineers to throw a bridge across the little stream. "Smed" drove over, and then said: "Now you men put that bridge under your arms and go back home. If you leave it here the Chinese will steal it." On Monday morning at my house when he wanted to return, he called up his chief of staff. I heard him say: "Colonel, I want the engineers to meet me at Yang-tsun at eleven o'clock with one bridge."

It amused me no end. Just like telling a man to meet you with a suitcase. Later he did a fine job for China on that same road. He told the Tientsin authorities that if they would furnish the labor he'd have his engineers supervise the building of a new road. The Chinese welcomed the suggestion and before the Third Brigade left Tientsin more than half the road had been completed. The part that the Marines built was by far the best road in China.

After Smedley Butler became a major-general and retired, he went into politics, which his friends deplored. He was always militant about governmental evils but was as tactless as a bull in a china shop. When a thing was wrong, it was obvious that it should be remedied. As a result, his hard-hitting, let-the-chips-fall-where-they-may policy made many enemies and he dashed his frail body against obstacles which only broke his health to bits and accomplished nothing. He died with the reputation of an incorrigible radical which he wasn't. For Smedley Butler, right was right and wrong was wrong. There couldn't be any middle line.

What with polo, social events, and magazine writing, the summer and autumn of 1927 passed quickly for Mac Young and me. We purchased a new lot of camels, got food and equipment packed, and cabled for the staff to come. As usual, we left April 17 and returned in September. The expedition of 1928 was successful in a new part of Inner Mongolia but ended in a distressing experience.

Politically the year had been most important. The Southern armies pushed rapidly northward. Chang Tso-lin's train was bombed by the Japanese and his troops withdrew to Manchuria. All China theoretically was under the Kuomintang, the "People's Party." A wave of intense nationalism swept the country; anti-foreignism was rampant. Because our expedition had received enormous publicity throughout the world, it was a spectacular target for several patriotic organizations, particularly one called the "Cultural Society."

When our caravan reached Kalgan from the Gobi, all the specimens were confiscated on the grounds that we had "trespassed on China's sovereign rights," "stolen the nation's priceless treasures," were "spies against the Chinese Government," and had been "searching for oil and minerals." This, in spite of the fact that since 1921 the expedition had been carrying on its work in co-operation with half a dozen Chinese scientific societies, and that the results were spread on the pages of every newspaper of the world. Our scientific Chinese friends did not dare protest for fear of being accused of pro-foreignism.

Eventually, the specimens were released but bitter feeling had been engendered on both sides. Negotiations extended to Colonel Stimson, secretary of state of the U.S.A., the Chinese minister at Washington, and the minister of foreign affairs of the new Nationalist government at Nanking. It was all senseless and absurd; we were simply the victims of circumstances. Any agitation of whatever character directed against foreigners found immediate popularity with the masses and their own government officials were powerless to resist. It marked the beginning of the end of foreign scientific work in China for several years.

Sven Hedin, the great Swedish explorer of Tibet, came to Peking to start an expedition in western Mongolia and Chinese Turkestan. He encountered the "Cultural Society." Sir Auriel Stein, one of the world's most distinguished Asiatic explorers, returned to England and the Royal Geographic Society after weeks of fruitless negotiations. He was vilified in the Chinese press in the most disgraceful way.

I was utterly disgusted by the time our specimens were released in 1928. Chinese scientific friends predicted that the agitation would end in a few months and urged us to try again. I did, and devoted the entire year of 1929 to costly and nerve wracking negotiations. As a result, by taking with us two poorly trained Chinese "scientists," giving to the Cultural Society half our valuable collections which were not even unpacked for years, we did one more expedition in 1930. It was restricted only to palaeontology for we were not allowed to make maps, and were so handicapped with stupid restrictions that our whole plan of operations was crippled. It simply wasn't good enough. I decided to quit even though we had only scratched the surface of the Gobi.

This unfortunate anti-foreignism in China stopped short a dream which had begun to develop in my mind just as did the Central Asiatic Expeditions. It would have been something bigger and vastly more important than anything we had done in our restricted explorations. My dream was to establish an "International Institute for Asiatic Research."

I am going to outline the plan briefly, at the risk of being a bore, for some day someone else will surely do it. Now that airplanes have demolished distances and the obstacles of land traveling, there are no longer unknown corners of the earth. My prophecy is that when the war is ended the world will enter upon a new and great era of intensive exploration. There are still vast little-known regions. Many of them are mapped poorly, if at all, and some hold undreamed of treasures in the realm of science. To study these areas; to reveal the history of their making; to learn what they can give for education, culture, and human welfare—that is the exploration of the future.

Most important of all is Central Asia. By that I mean Mongolia, southern Siberia, Chinese and Russian Turkestan, and Tibet. There is no other region on earth which will yield such important results in every branch of natural science. Central Asia is the home of rapidly vanishing peoples and of ancient civilizations that had a profound influence upon the history of the earth. If we are to understand our world of today we must know its past. The key lies in Central Asia. The scientific attack must be made systematically like the campaign of an army to insure best results. It must be international.

In addition to a president, field director, and other necessary officers, the international organization would have an advisory council represented by distinguished specialists in various branches of science, from all over the world. This body would decide upon the general plan of research, project the work to be undertaken each year, the scientists who would be invited to participate, and the ultimate dispositions of their collections. These would be deposited in various world institutions where they could be most advantageously studied and be accessible to the greatest number of students in that particular field. In return for such collections, an amount to be decided upon by the advisory council would be contributed to the endowment of the "Asiatic Institution."

The institution would not only finance each expedition but would arrange all details of permits, transportation, servants, equipment, etc. When the particular scientists arrived at the point of departure they could go into the field at once without loss of time or energy. These expeditions, necessarily requiring assistants of various kinds, would furnish an excellent training school for young scientists and explorers who now find great difficulty in entering the field of scientific exploration. Although the study of the collections would be made at the institutions where they were deposited, the scientific results would be published by the Asiatic Institution for world distribution.

China offers by far the most practicable place for field head-quarters. It is the gateway to Mongolia and southern Siberia, and all other parts of Central Asia can be reached from there with least difficulty. But first of all there must be the intelligent and active cooperation of the Chinese government.

Such was the plan I had conceived. I discussed it with scientific societies in England, France, Sweden, and Germany. Everyone was enthusiastic. I already had the pledge of a million dollars to start the endowment. All that remained was the official co-operation of the governments concerned. China's wave of anti-foreignism made it impossible. The chaotic state of Chinese politics, of course, was responsible and I believe that in due time, after the war, it will meet with the same enthusiastic response it had in Europe and America. It calls for a Utopian state of international co-operation, to be sure, but I do not think this is too much to expect in the future. Though it probably will not come about while I am alive.

OMG

Sir Hubert Wilkins has the same dream for an international meteorological institution where weather reports from stations all over the world can be assimilated and long range predictions made. It, too, will come true some day. Think what it would mean to farmers in widely separated regions if they could be told with accuracy six months in advance, where there would be drought and where excessive rainfall! Wilkins probably won't live to see it either, but he laid the foundations. Admiral Peary during the last years of his life was continually talking of airplanes in Arctic exploration. He did not live to see the day when Russian explorers landed at the North Pole and spent weeks drifting on the ice, learning secrets of the Arctic which have been of enormous value in the present war. But it was done. So, I am sure, will my International Institution for Asiatic Research come into being after I am dead.

Chapter 27

Fate Takes a Hand

For years up to 1932 I was a regular commuter between the Orient and America. Sometimes I went by way of the trans-Siberian railway, sometimes southward through the Suez Canal. Each trip around the world seemed to reduce the globe in size. There was no more mental or physical effort involved in starting for Paris, Vienna or Moscow, Singapore, Hong Kong or Peking than in taking a train to Chicago. I had a horror of accumulating possessions or responsibilities which might anchor me in any one spot. A small army trunk and a suitcase were sufficient baggage for I had regular stopping places all over the world and deposits of clothes. The ship lanes of the seven seas and the cities of Europe or Asia were as familiar to me as the streets of New York.

There were, of course, the usual rewards of a successful explorer: honorary degrees from universities, medals, and election to scientific societies. All of us on the expedition knew we had done a worthwhile job. The results spoke for themselves. The satisfaction and excitement I had had in conceiving, organizing, and directing the expeditions were the important rewards. Don't mistake my meaning. I liked receiving medals that in half a century had been given only to seven or eight explorers such as Peary, Nansen, Scott, Shackleton, Amundsen, Byrd. It was enormously satisfying to have my name enrolled with theirs. I'd have been less than human not to have found it so. What I am trying to say is that I considered the public recognition only as a pleasant side issue of a job which in itself had given me ample reward.

The personal devotion of every man, both native and foreign, who was a part of the expedition is the one thing I cherish above all else. The fact that during those years in the Gobi, in days of hardship and disappointment, sandstorms and sunshine, we never had a single quarrel is a great satisfaction. I know that each man of our staff thinks of his Gobi experience as among the most memorable of his life. All of them have told me so. We worked together as a happy family because each one showed himself to be worthy of the respect and affection of all the others.

Another reward was that I met hundreds of interesting people throughout the world. My house in China became a Mecca for visitors to Peking. With a perfect Number One boy and a score of servants, entertaining was as painless as mixing a cocktail. Almost everywhere I go now, someone will say, "I dined, or lunched, at your house in Peking." It was all part of a cosmopolitan life which couldn't have been more fun.

Along with the other things I made a lot of money lecturing and writing. Our expedition was spot news and magazine editors paid well. I did a series of articles on the field work for the *Saturday Evening Post* and six or seven more on adventures. These were combined in a book *Ends of the Earth* which sold like hot cakes. Always my lecture managers were pleading for more time. When I couldn't give it to them they boosted the fee. Even then I couldn't begin to fill the engagements offered. Money came easily and went just as easily. I never did have a grain of sense about my personal finances because wealth didn't interest me in the slightest. So long as I could do my job, that was all I cared about. If I had been sensible enough to put a third of my income into an annuity or a trust fund, I could have forgotten about money for the rest of my life. But, of course, I didn't. Some of my friends suggested that they could make me millions in the stock market so I let them "invest" my money. Then in 1929 it dissolved like mist before the sun. It was all gone; there wasn't anything! But it didn't make me lose a single hour's sleep. I couldn't visualize a time when wealth would be necessary to my personal happiness. It never will be. My wants and needs are simple. So long as I have a gun to shoot, a good fly rod, and work to do, what can money buy in the way of happiness? Nothing, so far as I am concerned. But the gentle spur of necessity is a fine antidote to mental stagnation.

The job ahead of me in the summers of 1931 and 1932 was to write the narrative volume of the Central Asiatic Expedition's final reports. Even though our work in the Gobi was prematurely ended the book could best be done in China away from the turmoil of New York. Out of the funds available for each year's exploration, I had set aside a part for publication. No other big expedition has ever done that. Usually at the end of their explorations they have to shop around to get money for publication and it is exceedingly difficult. Field work is soon forgotten. Only the publications remain as a permanent record of achievement. At the end of my first expedition the results began to appear in short papers. New species were described; new facts put on record. A hundred and forty-five have been published. A series of twelve final volumes were projected under the general title *The Natural History of Central Asia*. Seven already are completed and I look at them with pride. As long as science exists they will stand as the "Systema Natura" of Central Asia.

Writing the narrative volume became the real point of my life. Dr. Sven Hedin, the great Swedish explorer, happened to be doing a book at the same time and Dr. Davidson Black was studying the remains of the "Peking Man," a primitive human discovered in the Western Hills. All three of us found we could work best at night. I played polo in the afternoon, got physically relaxed, and had a light dinner. A pile of sandwiches and a bottle of beer were placed on a tray in my office. As the multiple activities of the day ceased, the compound became quiet and only the calls of street vendors sounded beyond the walls. Then I settled down to write.

Usually about three o'clock in the morning I telephoned Hedin and Black. We would meet at the Alcazar or International, night cafés of somewhat dubious reputation, have scrambled eggs, dance with the Russian girls, and then go home to bed. I woke only in time to bathe, have luncheon about three o'clock, and ride or play polo. Social activities for me were out. No luncheons, dinners, or cocktail parties. Peking knew me no more as a host. I was a night worker and enjoyed it. By the end of the summer of 1932, the book, *The New Conquest of Central Asia,* was finished. It is six hundred and seventy-eight large pages.

No longer was there reason to maintain a home in China. For twelve years I had lived in the great house at No. 2 Kung-hsien Hutung. It had become something personal to me, a part of my very self. When for the last times I walked through the big red gates, tears were in my eyes.

In New York I made a new home for myself at the Hotel des Artistes, No. 1 West 67th Street. On the roof, long after the famous building was completed, they built a little house. Its front door opened onto a lovely terrace, where I planted hundreds of flowers and some small trees and installed a fountain in the rock garden. Looking across the Park to Fifty-ninth Street the view was lovely in the daytime and dazzling at night. With some of my treasures from Peking I made it into a perfect little Chinese home.

In the block on Sixty-seventh Street, between Central Park West and Columbus Avenue, lived more artists and writers of international reputation than in any other part of New York. Just below me was the famous illustrator Wallace Morgan; also Howard Chandler Christy, Heywood Broun, Mrs. William Brown Meloney, and Neysa McMein. Across the street Burton Holmes had a beautiful apartment; a few doors down lived Fannie Hurst, Will Beebe, and James Montgomery Flagg. At any time of the day or night some hospitable door, where one was sure to find interesting company, was open for we all lived a Bohemian existence.

A short time after taking the penthouse I had my only adventure in New York with a burglar. An iron ladder led up the wall from Wallace Morgan's terrace to a tiny balcony opening out of my bedroom. I slept about six feet from the French windows of the balcony. Years in the field have made me very sensitive to anyone in my room. Just before dawn, I raised on my right elbow, wide awake. A man was crouching beside the bed. A regular stage burglar; soft cap pulled down over his eyes and a short coat tightly buttoned. There he sat looking at me, evidently waiting to see if I were asleep before he began to prowl. For an instant I thought it was a dream. Then a wave of indignation swept over me. "You —— " I yelled and jumped at him.

Apparently that was the proper way to address burglars for he leaped backward and made for the balcony. My feet got tangled in the sheets and I landed on the floor on all fours. The burglar was half over the iron rail when I caught him by the leg with one hand. He kicked like a mule. The first one landed on my chest; the next got me full in the face. He tore loose, ran down the ladder like a monkey, and through the terrace door into the hallway.

I phoned the night operator at the front entrance and he called the West Sixty-eighth Street Police Station. In less than five minutes seven policemen were searching the twenty-story building, but the burglar got out through the basement fire door. It was gray dawn and the man's silhouette against the sky made me sure he was an Oriental. There had been several newspaper articles about the Chinese art treasures in my apartment; doubtless that was the reason for his visit.

I could have been happy up there in the sky in my eagle's nest but things were not as they should be in my mind. The trustees created for me the post of vice-director of the Museum. But the director, Dr. George H. Sherwood, one of my oldest and dearest friends, was the sort of man who must do everything himself. The assistant director, Wayne Faunce, he had trained in his way of work. There was nothing important for me to do. For the first time in my life I had no objective, no goal. I was definitely unhappy. There were hundreds of friends in New York and hostesses of the "smart set" took me up in a big way. My engagement book was full for weeks ahead. Of course, in the innocence of my heart I thought my "social success" was because I was so completely charming. Actually, I was only an eligible bachelor; just an extra man, a "filler in."

My problem came near to being solved temporarily by Douglas Fairbanks, Sr. One day the phone rang. "Roy, Mary and I are at the River Club. Can you come over right away? I've got something exciting to talk to you about." When I got there, Douglas was striding about the sitting room.

"Ever since I went to China," he said, "I've wanted to do a picture out there. But I couldn't find a plot. I've got one now. You know Madame Wellington Koo? She and I thought up a story and I turned it over to a script writer. I've just read it. It's a honey. Sort of a synopsis of big events in Chinese history, ending up with the present day. Everything centers about a young Chinese airman who saves the country in the midst of a war. Of course, we won't say so but it's war with Japan. You and I know that's bound to come soon. Since the Japs took Manchuria last year they'll push southward just as soon as they think they can get away with it. This picture will not only be damned good entertainment, but it may help the world to see what the Japs are doing.

"I want you to help me. Of course, most of it will be done in Hollywood but some must be shot in China. Your knowledge of the country and people could be of enormous use. You can be 'assistant producer' or something of that sort. I'll give you a thousand dollars a week. If that isn't all right, tell me what you want. Will you do it?"

I couldn't think of anything I'd rather do. Working with Douglas Fairbanks would be a lot of fun and I'd have a chance to see the inside of the movie business.

"I'll be ready in about three months," he said. "I've got to do another picture first. Then the decks will be clear. After the picture is done let's you and me have that tiger shoot we didn't get. I'm never going to be happy until I kill one of those long-haired tigers."

At this time, Fairbanks was in a phase of frenzied traveling. He never wanted to stay put. I happened to mention that during a quarter of a century I had not spent an entire year in any one country. He jumped to his feet. "I want Mary to hear that. Mary, Mary, come out here." The bedroom door opened and Mary Pickford came in. We were introduced. "Roy says he hasn't spent twelve months in any country for twenty-five years. What do you think of that?"

"I think," said Mary, smiling wryly, "that he will be a bad influence on you. I don't know him but I definitely do not approve." With that she shut the door. I guessed that his restless dashing about the world was a sore point in the Fairbanks ménage.

Douglas went back to Hollywood. He sent me a dozen long telegrams for he never could take time to write. Then I had a wire to meet him in New York. His son, Douglas, Jr., was ill in Paris and he was sailing the next day. Other unexpected things came up, I've forgotten what they were, and he never made the Chinese picture. I was awfully disappointed.

After that first winter in New York I could not face a summer of aimless existence. "Europe for me," said I. "I'll do a spot of wandering. I've always been happy moving about. That is what I need."

I'd never seen the château region of France, so I cabled for a car and chauffeur to meet me at Cherbourg and sailed on the *Berengaria*. But France didn't do a thing for my restless spirit. No sooner had I arrived in one place than I wanted to be off to another. "It's too damned banal," I thought, "too many tourists. The Balkans, now, will be different."

So I went to Yugoslavia, Zagreb, and the lovely lake at Bled, the summer capital. It was just the same.

I went to the Austrian Tyrol. When I had walked across the Brenner Pass twenty-three years before I had had a wonderful time, but now it left me cold.

Then to Berlin where Hitler was beginning to change the face of Germany. No interest. One night sitting alone at a table in a Berlin beer hall, disconsolate, I got a little drunk. Just enough for me to face facts and see myself clearly.

"What you are trying to do," said I to my glass of beer, "is to run away from yourself. But you can't do it. Yourself is always with you. For the first time in your life you aren't heading anywhere. You are an explorer. For twenty-five years you've lived in the field and you can't be happy in the conventional life of a city. It doesn't fit. Go back to the desert where you belong."

I sat in that beer hall, thinking, until three o'clock in the morning. When I returned to the hotel I was mentally at peace for the first time in six months. Another expedition was in my mind. Russian Turkestan! It would have to be a co-operative show with the Soviets. We'd work there in the summer and Iran in the winter. I'd stay for five years—never come out. To hell with cities!

The next day I went to the Russian Embassy, got my visa and letters from the American ambassador. Then took a plane to Leningrad to see Dr. Borissiak, president of the Russian Academy of Sciences. I found him vacationing at a "co-operative country home"—a pretty grim place for a distinguished scientist. We never had met but he kissed me on both cheeks like an old friend. For hours we talked; all that day and the next and the expedition took shape. Then to Moscow to see Karakhan, the Assistant Commissar for Foreign Affairs, whom I knew from Peking where he had been ambassador. It was all arranged. I'd furnish the money, the Soviets would give material aid, and we'd split the collections, half and half. I could hardly get back to New York fast enough. Suddenly all my vague unhappiness was gone for I had a new and thrilling job.

Professor Osborn had reached his twenty-fifth anniversary as president of the board of trustees of the American Museum of Natural History. He wanted to resign and my friend F. Trubee Davison agreed to accept the post. Trubee had been Assistant Secretary of War for Air in President Hoover's cabinet. Roosevelt was in and Hoover out. Trubee was temporarily jobless. No one could have been better fitted to fill that particular post at that particular time than Trubee Davison. Madison Square Garden wouldn't have held all his friends in New York City alone. He knew everybody and everybody liked him, for his warm affectionate nature drew people to him like a magnet.

I couldn't have been more delighted than when Trubee became president of the Museum. I knew he would give me the same enthusiastic co-operation in my new plan that I had always had from Professor Osborn. A radio program for the winter would net twenty-six thousand dollars. This money I would turn in to the expedition funds, and in spite of the Depression the rest was sure to come. I set about the money raising program without delay.

Then a turn of fate changed my whole life. I was sitting in my office when a secretary rushed in to say that the director, Dr. Sherwood, was very ill. He was lying in his chair gasping for breath. It was a heart attack, I knew. The doctors said he must have six months' complete rest. Trubee was sunk. Not only was he very fond of George Sherwood, but it left him in a difficult spot. "I've just taken over this job, the director is out, and you are going away. What a mess!"

"If you want me to, of course, I'll carry on until Sherwood gets better or you find a new director. The expedition can wait," I said. He was relieved and the trustees asked me to become acting director for six months. It was something interesting and constructive. All my restlessness vanished.

I enjoyed myself enormously that summer for there was a job of reorganization that needed to be done. Trubee and I worked together perfectly because our minds functioned the same way. He had complete confidence in me and I not only admired his ability but developed a deep affection for him which has grown with the years. Of all men I have ever known I would wish to be like Trubee Davison.

The Museum work was confining but I kept fairly fit by riding every day in the park. I couldn't give up horses so I joined the Fairfield (Connecticut) Hunt Club, bought a string of ponies, and played polo twice a week. Then one day in August I got an awful spill. Two lumbar vertebrae were badly injured. The doctors said that another fall would either kill me outright or I'd be a cripple for life. I must give up riding. That was like a sentence of imprisonment, for my athletic interests had centered about horses for twenty years—polo, fox hunting, steeplechasing. I didn't know what to do.

In November, it became evident Dr. Sherwood could not resume his place as director. The trustees asked me if I would accept the post but in spite of the fact that I enjoyed working with Trubee intensely, I didn't want the job. My heart lay in the desert and I hoped soon to go back to the sunsets and the sandstorms. But the trustees said I need not give up exploration. Perhaps the expeditions could not be so long but they would wish me to continue field work. I still hesitated. That I should become director of the institution where I had spent all my life was a logical conclusion but I could not be certain it was best for the Museum or for me.

Then on Wednesday, an intimate friend, Dr. William H. Holden, telephoned. "I haven't seen you for some time. The most beautiful girl I've ever looked at is coming to my office this noon. Why don't you come and we'll lunch together?"

"All right," said I. "Can do."

I had been in Bill's office about ten minutes when the girl walked in. She wore a mink coat and a little fur hat over a crown of golden hair. Her eyes were tawny, but a moment later I decided they were green. No, not green, gray. We were introduced. "Luncheon at the Museum," I said, "is indicated. I think you'll enjoy it." She did. For two hours we walked about the halls and into the department of preparation, where all sort of interesting things were going on. Then she and Bill left.

I couldn't get her out of my mind so I asked her to dine with me Saturday evening. We talked from seven o'clock until two in the morning. Her name, by the way, was Wilhelmina Christmas, but no one ever called her that. She was Billie to all her friends. I begged for the next evening, but she couldn't so it had to be Monday. All day Sunday I was restless as a fish out of water. I telephoned her twice but her maid said she wasn't home. Monday night finally came. Over cocktails I said, "You are going to marry me. I don't know when, but it is inevitable. You might as well make up your mind to it."

On Tuesday, Trubee asked me again if I would not take the directorship of the Museum. "Certainly. I'd love to. I'm going to get married." He gasped. "But only last week you told me you never would marry again!"

"Of course I did, but that was last week. This week it's different."

He only shook his head. "You work too fast for me," said he.

It was as great a surprise to me as to him. I had built up a philosophy of life in which marriage never again would figure. It was all very clear and logical until I met Billie; then the theory collapsed of its own weight. Suffice it to say that we were married three months later. But it wasn't as simple as it sounds.

Jo and Anne Sheedy, intimate friends of Billie's, asked us to be married at their apartment on Park Avenue. They had a little girl, then aged five. When the newspapers announced our coming marriage, the Sheedys received a note saying that unless they sent a large sum of money immediately, the child would be kidnaped. Jo turned the letter over to the police. Then another came stating that if the wedding went on, the place would be bombed.

It was all very hectic. Every package that came to the Sheedy home had to be sent to the police department to be X-rayed, and detectives shadowed the child whenever she moved from the house. The strain became too great. The Sheedys packed up and went to Europe. Billie and I decided to be married quietly a day earlier, on February 21, and the police wished it to be kept secret. The ceremony was supposed to be on February 22 as announced and at that time they hoped to catch the would-be kidnapers. We were married on Thursday and sailed Saturday for the Pacific coast via Panama.

At Havana, the ship's first stop, we took a motor at the dock for a drive about the city. Suddenly the chauffeur swerved around a corner on two wheels nearly throwing us out of the car. "A bomb. A bomb. There in the street, in front of the President's palace." Sure enough, we could see a black object smoking on the pavement. It was tear gas. Billie burst out laughing. "We attract bombs like magnets. What next?"

There weren't any more bombs, however, and after two months we returned to live in a duplex apartment at the top of the house belonging to the late Joseph Pulitzer, famous owner of the *New York World*. After his death, Ralph Pulitzer converted the palatial residence into apartments. Because Mr. Pulitzer was blind and extremely sensitive to noise, the entire house had been sound-proofed. With closed windows it was as silent as the country.

Billie and I converted the apartment into a lovely Chinese home, red woodwork, gold ceilings, and walls covered with Chinese embroideries. There was, too, a wide terrace which became an old-fashioned garden on a New York rooftop. The care of plants, apple trees, and flowering shrubs kept us busy on Sundays and holidays. Over the surrounding houses we could glimpse the traffic on Fifth Avenue and the green of Central Park; otherwise the seclusion was more complete than one would dream possible in the heart of the world's greatest city. Breakfast in the summer was always served in the garden and we dined there at night. My friends said they thought I was very lucky. I knew damned well I was.

Chapter 28

Dangling in the Depths

A few months after returning from the Pacific coast, Billie and I went to Bermuda where I had one of the strangest experiences of my life. It was in a diving helmet, dangling on a thirty-five foot ladder over the side of a bobbing launch in deep water off Nonsuch Island! Will Beebe, our host, said it would be amazing and I believed him academically, but actually I had no conception of what I was about to feel.

First we went down off shore amid the gorgeous growth of sponge and coral, under the eaves of a vast shelf of rock. Billie was totally unafraid, reveling in the exquisite beauty of the watery world about her. There we could see the surface ceiling, bright amber and gold from the sun above.

The next dive was on a dark day in a sheltered lagoon beside an old wreck. Through the water glass I watched my wife walk about on the sand among the coral hillocks. The squirrel fish, blue heads, striped sergeant-majors, and dozens of other fish clustered about her slender body as she wandered with one hand extended like a sleep walker among the water flowers of nature's most lovely garden. She went close to the prow of the sunken vessel and there knelt down, utterly absorbed in the magic of the rainbow-colored sponges, sea fans, and delicate anemones which clothed the oaken timbers in robes of splendor. I had a thrill of fear as she drifted around the hulk and was lost to sight, drawn on and on by the wonders of what to her, and me, was a new and undreamed of world.

Then Will headed the boat for the restless water outside the shelter. A lowering, rainswept sky shut down close over the choppy sea. He dropped the ladder and descended. "It's wonderful. There's a regular bathysphere blue in the water," he reported.

Then I went down. At first I did not try to look. I was too occupied in staying on the swaying ladder. Near the end I stopped, clung with one hand and let my body float off obliquely. I was enveloped in a strange intense moonlight blue, darkening imperceptibly. Below, the same weird blue. I looked up. No water ceiling, no comforting shadow of the boat; only blue and darker blue. The absolute silence was appalling. A few feet away three great circular jelly fish, ghostly white, floated past. Others rose almost under my swaying body; a single colorless pulsing mass hung like a halo just above my head.

Suddenly I was afraid. A nameless terror took possession of my bones and flesh and blood. It seemed that I had died and gone to some strange place unknown to human minds! Corpses, dead white, women with streaming hair floating past, staring at me with sightless eyes, should have been part of that ghastly world. The only touch of reality, my one tangible hold on the life I knew, was the rung of that iron ladder.

Rising to the surface was like waking from a nightmare. Yet we had not reached shore before I wanted to go down again. Analyzing, unemotionally, what had happened, I realized that I had been so completely absorbed in sensations that my brain had ceased to function. Had there been anything familiar, fish that I knew, coral, something other than that vast weird emptiness of blue and the ghostly jelly fish, I would not have been affected so profoundly. Doubtless were I to drift down again in the same place and under the same lowering sky I would not have that sensation of bodily detachment. But I am glad it happened just as it did. Otherwise I should have missed an extraordinary experience.

After Bermuda we stuck close to the Museum until the following March when Billie and I sailed for China. I wanted to show her something of my Oriental past and yet I was a little afraid. Would the sight of my abandoned palace and Peking spoil a wonderful memory? There would be new people whom I didn't know and a new life. I've always thought it a great mistake to revisit places where one has been very happy. Subconsciously one expects to find everything exactly as when one left and there is usually bitter disappointment. But I decided to take a chance and by cable rented a house in Peking.

After a delightful fortnight in Japan, we sailed from Kobe for Tientsin. A dust storm ushered us through Peking's Tartar Walls where my old servants were waiting with flowers in their hands. Then into a car and up the Nan Chi-tzu to the Street of the Small Market. We were hot and tired and very dirty. As the gate to the compound opened, the clink of ice in a cocktail shaker sounded. Heavenly music! The boys were old friends and a diminutive *amah* had a hot bath ready drawn for Billie. It was a typical Peking reception.

Next day, out in the city, my fear of coming back ended. I was in Peking again, the same sights and pungent smells, the same street calls, the same pigeons with whistles on their tails circling above the yellow roofs of the Forbidden City! Everything was there but I seemed to have known it only in a previous existence. My house on the Bowstring Street was sadly in decay. No one had lived in it since I left. As we roamed through the courtyards and the vast rooms, thick with dust, I thought of myself in the past, in the third person. I was like an embodied ghost. The past was as impersonal as though I had been a character of history.

Peking was much improved. Some of the worst streets had been paved and a few of the temples repaired. But an atmosphere of impending doom hung over the city. In spite of diplomatic protests the Japanese were pouring troops and munitions into China. Their swaggering soldiers pushed foreigners off the sidewalks and spewed insulting epithets when a white girl passed. Every morning if the wind was right one could hear the rattle of machine-gun fire from beyond the Tartar Walls. East Sunit Wang, the most powerful Prince of Inner Mongolia, an old friend, came to see me one day. He had been summoned peremptorily to the Japanese Embassy. In blunt language they had said, "Are you with us or against us? If you throw in your lot with Japan your lands will be free, your people untouched. If not, you will be dead." After all he had little choice. He thought the Chinese powerless to resist Japan. This was less than a year before the "incident" at the Marco Polo Bridge when the spark was struck that started the invasion of China.

Like all others who knew the Orient, I was sure Japan would lift the curtain when the stage was set. We were glad to leave for America before the first act began.

Back in New York, we made a few trips to Texas and Florida but the Museum work became continually more confining. Still I thoroughly enjoyed it. The trustees were all my personal friends and a splendid group of men. Our apartment was a meeting place where the really important questions were settled over a cocktail or Scotch and soda. The only fly in the ointment was my lack of fresh air and exercise. A club gymnasium wasn't much use. Exercise never did me any good unless it was in the open and I enjoyed it. Sitting on a rowing machine or playing handball bored me. The confinement and nervous strain began to tell. My days were full of people—talking, talking, talking. Weekends, Billie and I tried to find some place where we could be completely alone. For twenty-five years I had lived in the field, sleeping in the open, and I was like a wild animal suddenly put in a cage. Never in my life had I been ill, but then every germ in New York City found me a delightful place for breeding purposes.

On Memorial Day, 1937, Billie and I were motoring back from a weekend on Long Island. We saw signs along the road, "Acre for sale." I said, "How wonderful if we could buy a little place in the woods, put up a prefabricated house, and spend weekends there where no one would know where we were."

"Yes," she said, "it's an idea, but not Long Island. Connecticut!" The following day Billie called a friend, Louise Lundy, who deals in real estate. "I've just the place," said she. "Only it isn't one acre; it's a hundred and fifty and there's a house and a pond."

Billie lunched with her next day, Wednesday. On Saturday we motored out to see the property. Actually we didn't know where we were going. We only followed the route numbers. They led us to the northwestern corner of Connecticut to the tiny village of Colebrook, with a church, a general store, two or three white houses, and a post office. We were charmed instantly. It might have been a living picture of Colonial times. Old drawings and photographs show that actually it has remained virtually unchanged since the days of the Revolution. Three miles away was Pondwood Farm. We didn't know where we were and didn't care after we saw the place. A small house, not very good, looked out to a beautiful pond set like a jewel in the forest. Pines and maples, silver birch and great oak trees bordered the water and at the foot of the hill was a perfect trout stream. We walked over the property, entranced. That was Saturday.

On Monday, Billie owned Pondwood Farm. It was sadly run down, yet that was a virtue in our eyes. Tangles of underbrush grew everywhere but we could cut them out as we wished and make it our own place.

The Tuesday after we became owners I met William Chenery, publisher of *Collier's* magazine, at the Dutch Treat Club. "Bill, I've become a farmer," I said.

"Where?" said he.

"Near a little village in Connecticut; probably you never heard of it. Colebrook."

"Well, I've a place three miles from there." Charlie Colebaugh, editor of *Collier's,* had a country house close to Bill Chenery's, and so it went. Jack MacMurray, formerly minister to China and one of my oldest friends, lived in summer at a camp on Doolittle Lake two miles away; ex-Senator Fred Walcott was at Norfolk, six miles from Colebrook. All of them were men I knew well and liked immensely.

Though Billie and I had bought Pondwood Farm overnight, without looking at another place, we could not have found another property that would have suited us more completely had we searched for ten years. Groves of pines and hemlock, mixed forest, pond and swamp held birds of every kind. Splendid fishing and shooting were at our very doors. Again my Lucky Star had guided me, and together we had followed it blindly to happiness.

The story of how we stocked the pond with bass, dug a swimming pool from a depression in the forest, converted an abandoned cellar into a lovely rock garden with an open fire place, would be a bore. To us it was fascinating and exciting. For the first time in our lives we owned some of the "good earth."

Never before had I wanted possessions; things that would anchor me to one spot. I wished only to be free to leave at a moment's notice for the ends of the earth. Billie, too, had the travel fever. Now we didn't want to go anywhere. It was, I suppose, the natural corollary to a life of continual change, like the desire of every deep sea sailor to have a farm and raise chickens.

At first we spent only weekends and a few months of summer at Pondwood Farm, using oil lamps for light and fireplaces for heat. Then one day Billie decided to build. I've found the best way to keep a wife contented is to let her either travel or build. Neither of us wanted to travel then so we built. For days, Billie pored over plans with an architect. She has much more foresight than I in our personal affairs.

"This," said she, "is where we will live eventually so we'll make it permanent."

I smiled indulgently and let her have her way. How right she was! The result became a living room with a fifteen foot plate glass window overlooking the pond, a pine paneled gun room, an oil furnace, and the last word in a kitchen and bathrooms.

For my part, I built a big log cabin just within the woods, completely concealed from the house. That is where I write. It seems to be almost a part of the forest. Indeed it is so much so that one day while I was sitting at my desk facing the open door a buck deer walked half into the room before he caught my scent. Rabbits, squirrels, and chipmunks visit me often, hopping about the floor if I sit very still. Just above the door a phoebe bird nests every spring and in a dead oak tree a huge pileated woodpecker buries acorns for future use.

Pondwood Farm with its fresh air and exercise and the multiple interests of nature made me begin to live again. We motored the one hundred and twenty-five miles every Friday afternoon and back on Monday morning. It seemed the solution to the confinement of city life.

Things of great interest were happening in the Museum. The African Hall, the dream of Carl Akeley, was nearing completion. Akeley had died in Africa before the hall itself was built, but he had planned every detail in a huge scale model. After "Ake's" death, Daniel E. Pomeroy, one of the trustees, made it his business to see that the hall was completed as a memorial to his friend. Nothing like it exists in all the world. It is Africa. Not only the animals but the trees, the leaves and grass, the very earth itself, were brought from the place where each group was collected. As a rule, it is futile to say that a thing never can be done better. But I am perfectly willing to make that statement about the African Hall. Bigger groups can be built, but not better ones.

October 1935 saw the completion of the Hayden Planetarium as a part of the American Museum. It was a great personal triumph for Trubee Davison. He persuaded the Reconstruction Finance Corporation to advance the money for the building but the instrument, made in Germany, they could not buy with U.S. government funds.

Trubee went to Charles Hayden. "I don't quite know why I'm here," he said, "but I heard you were interested in planetariums. We need a hundred and fifty thousand dollars for the instrument in our new building. Perhaps you'd care to help."

Charles Hayden looked out of the window for a moment. Then he turned to Trubee. "Yes. I am interested in planetariums. Never have I been so affected as when I saw the one in Chicago. It is, I think, a great spiritual force in that city. We need it in New York. I'd be delighted to give you the hundred and fifty thousand. We are all a little vain; I hope it may bear my name."

That is, substantially, the way it happened. Trubee was out of Charlie Hayden's office in five minutes and the planetarium was an established fact.

The night before the public opening of the planetarium, Hayden gave a dinner at the Waldorf for his personal friends; about a hundred and fifty of them. A private showing at the planetarium came after dinner and most of his guests arranged "aftermath" parties. Billie asked Junius Morgan and his wife, the Davisons, and several other trustees and friends to come to our apartment for a drink. But strangely enough, no one had thought of our host. He was left completely at loose ends. We went home a little early. The doorbell rang and there was Charlie Hayden. "No one asked me anywhere," he said, in a very small voice, "so I thought I'd come here." Just then Billie came down the stairs. Although Charlie hardly knew her, he threw both arms about her neck and kissed her. "I'm just so happy I've got to kiss someone. This is one of the biggest nights of my life." He was just like a child, even though he was a world-famous financier, director in a score of companies that controlled much of the wealth of the United States. When he died Charlie willed fifty million dollars to a foundation for the "Youth of America." How any man who had lived in the exact commitments of high finance could have left such a nebulous bequest is a mystery to me.

The night after Charlie Hayden's private opening, the planetarium had its first public presentation. We were all there again, of course. One of the speakers of the evening was Robert Moses, New York's brilliant commissioner of parks. No one can predict what Bob Moses will say at any time. He ran true to form that night. During his address he remarked with a smile, "Charles Hayden has purchased immortality for a hundred and fifty thousand dollars. That's cheaper than anyone ever got it before."

Plans were under way for a Hall of North American Mammals somewhat like the African Hall. Trubee Davison threw himself enthusiastically into bringing the exhibit into being. He appointed one of the trustees, Robert E. McConnell, chairman of the committee. Bob did a fine job.

The City of New York was obligated to build the hall and provide the cases; the trustees to finance the exhibits. Friends had contributed nearly a quarter of a million dollars in groups but the hall was incomplete. We needed only fifteen thousand dollars to finish the central section and make the hall available to millions of visitors. Otherwise the whole place must be closed indefinitely. The city was feeling very poor. Mayor La Guardia was particularly concerned about finances. Although the director of the budget and the comptroller approved appropriating the fifteen thousand, the mayor turned it down. Lewis Douglas, one of our trustees, argued with him to no avail. Others tried to make him see the light, but he was adamant. Too many schools and hospitals, he said, needed the money urgently. The North American Hall would have to wait. It didn't make sense. We all felt sure that if only he could see the place he would realize it was poor business to tie up a million dollar hall for the want of fifteen thousand dollars, but he wouldn't come to the Museum. Finally I got the mayor in the hall by a trick. I can't resist telling the story here and I'm sure the mayor will forgive me.

On a Friday the mayor's secretary telephoned that His Honor wanted to come to the planetarium on Saturday afternoon with his little son, Eric. It was, however, to be an entirely unofficial visit; he was coming for relaxation. An evil thought came to my mind. Could I not entice him into the North American Hall? Eric was the solution. We had a group of Alaskan brown bears almost completed and one of them stood erect with arms outstretched. What boy could resist the biggest bear in the world?

When the mayor arrived, I met him at the planetarium entrance. We had been acquaintances for years and since this was an unofficial visit I scrupulously addressed him as "Major." Eric and a friend were with him but my heart sank when the Mayor said, "How long does the show take? I've got to get back to City Hall for an important telephone call from Washington." There, I thought, goes my plan. But my Lucky Star took charge. Just after the performance, an attendant told me that the Washington call for the mayor was on the wire; he could take it in the curator's office. While the mayor was talking, I said to Eric, "Wouldn't you like to see the biggest bear in the world? You can even touch him."

"Oh, gee, *yes*. Can I really *touch* him, Dr. Andrews?"

"Yes, you can but you've got to persuade your father to wait. If you want to see the bear it's up to you."

When the mayor came out of Dr. Fisher's office he said, "Well, Eric, we've got to be getting back. Come along."

"But, Dad, Dr. Andrews said I could see the biggest bear in the world. He said I could touch him. I don't want to go till I've seen it."

The mayor turned to me. "Where is this bear Eric is talking about?"

"Just a few steps from here. It will take only a moment."

So we went to the tightly closed North American Hall. At the end stood the great bear. Eric and his friend raced for it like a couple of colts in pasture. The mayor gave me a quizzical look. "What hall is this?"

"Why," I said innocently, "it's the new Hall of North American Mammals."

"Oh, *that's* the reason you wanted to show Eric the biggest bear in the world! This is what Lew Douglas and all those other fellows have been talking to me about. Well, now I'm here, let's see it."

"But," I said, "this is an unofficial visit. I promised not to talk city business."

"Oh, forget that. Tell me about it."

The mayor was impressed as I knew he would be.

"If I give you that fifteen thousand dollars will you promise not to ask for any more money?"

"I'll promise not to ask for any more *this* year," said I.

That was Saturday. On Monday I told Trubee the story and showed him a letter I wanted to send the mayor. I was counting on his sense of humor. Here is a copy of the letter.

PERSONAL

March the third

Nineteen hundred forty-one

Dear Major La Guardia:

You have seen the "biggest bear in the world," and so has Eric, but no one else of the two million people who visit the Museum each year can see it unless you are able to persuade the Mayor to give us $15,000 to complete and open the central section of the North American Hall. You might tell the Mayor that, as Director of the Museum, I promise not to ask the City for any more money this year, if he will give us the $15,000.

I do hope you can convince the Mayor that it is good business to spend $15,000 in order to make a million dollar hall available to the public right away.

With kindest regards,

Faithfully yours,

(signed) Roy Chapman Andrews

Director

R.S.V.P.

Major Fiorello H. La Guardia

City Hall

New York City

It got results at the City Hall and Bob Moses's office immediately. They said we would get the fifteen thousand. Four days later the following letter arrived from the mayor.

PERSONAL

CITY OF NEW YORK

OFFICE OF THE MAYOR

March 7, 1941

Dr. Roy Chapman Andrews

The American Museum of Natural History

New York City

Dear Dr. Andrews:

I am sending on an authorization for $15,000 to complete the Central Section of the North American Hall. As soon as the city is in a financial position to spend more money on the Museum of Natural History, I will avail myself of making another private visit to the Museum to see many things I wanted to see but which I now necessarily must defer.

I now claim the distinction of having paid the highest admission fee in the history of the Museum for looking at the biggest bear in the world.

Very truly yours,

(signed) F. LA GUARDIA

Mayor

I replied as follows:

PERSONAL

March the twelfth

Nineteen hundred forty-one

Dear Mr. Mayor:

We are all delighted about the recommendation for the Central Section of the North American Hall.

I am sorry that the admission price was so high but after all I think it is going to be worth it. Anyway, since you have paid your fee for the year I wish you would come again and I will promise to show you and Eric the new exhibits without costing you a penny, whether you come as Major La Guardia or as the Mayor.

I do want to tell you again how greatly we appreciate your action.

With kindest regards,

Faithfully yours,

(signed) Roy Chapman Andrews

Director

Honorable F. H. La Guardia

Mayor of the City of New York

City Hall

New York, N.Y.

The cases were completed and the hall opened. Millions of people have enjoyed its beauty.

Chapter 29

A Square Peg in a Round Hole

There was a lot of satisfaction in my work during those first winters. I was doing a museum man's job where results showed in exhibition or scientific work. But the institution began to feel the pinch of finances more every year. Drastic curtailments were necessary. It got to the point where, with a million and a half dollar budget, I was trying to save fifty dollars on paper drinking cups and towels; where when a collection worth ten thousand dollars was presented to us, we couldn't take it because we didn't have three hundred dollars to pay the cost of packing and transportation. Appropriation for scientific publication was cut in half. I had to write letters of dismissal or retirement to many of my oldest friends. It was no one's fault. The trustees were doing all they could but we just didn't have the money to adequately finance the institution in the face of rising costs and reduced income. Something had to be done.

A campaign was inaugurated to raise ten million dollars for the Museum. From that time on, my job changed completely. I was a museum director in name only; actually, merely a promoter. I had promoted my own expeditions successfully but that was only for a short period of individual high pressure work with the romance of exploration to sell. The knowledge that field work was waiting as soon as the distasteful business of money raising ended made it endurable. Now the possibility of continuing exploration was gone for so long as I remained director.

Many of New York's influential people—I might say most of them—were unfamiliar with the American Museum. So a plan was devised to have prominent New York hostesses give dinner parties at the Museum in the beautiful Portrait Room of the Roosevelt Memorial. Sherry's catered and the dinner was always excellent. We provided the entertainment, taking the guests about certain halls and showing them the real inside workings of a great museum. The hostesses paid the bills. About forty dinners were given during the winter and they became a feature of the New York social season.

Both Billie and I enjoyed them for a time, but three or four a week became pretty hard to take. Still, we met interesting people and that was pleasant. As one of the world's greatest institutions every prominent visitor to the city was brought to the American Museum. After Al Smith had shown them the Empire State Building they came to us. Of course, being director, I did the honors.

I remember that Prince Chichibu, brother of the Japanese emperor, Hirohito, stopped on his way to England as Japan's official representative at the coronation of George VI. He had only two days in New York but we were notified weeks in advance that a visit to the American Museum was a great desideratum of His Imperial Highness. At nine o'clock in the morning Trubee and I met him at the door of the Roosevelt Memorial. He was a wooden-faced little man, very correct, followed by half a dozen morning-coated hissing Japs. First we conducted them to the planetarium. The Japanese national anthem was played. We all stood at attention. Then the stars were shown as they would appear that night over Tokyo. For three hours we guided the prince about the halls. My body and brain were exhausted but he was going strong at the bitter end.

"Your Imperial Highness must be tired," I said.

"Oh no, we are used to this. We are trained to it."

"What are these?" he asked as we passed through the Vernay East Asiatic Hall and stopped at a case of rare Indian lions.

Trubee swears that I replied, "These, Your Highness, are African tigers." But I'm sure I didn't say any such thing.

For some reason, among the hundreds of famous people who visited the Museum, my memory harks back to the late Dr. Dafoe, the "deliverer" of the Dionne quintuplets. Perhaps it is because he was such a simple, charming man. The keys of the city were virtually handed over to him. Manhattan set out to give him a good time and if he didn't have it at least the city did. It was reported in the papers that the doctor most wanted to see the American Museum of Natural History. So I telephoned an invitation to visit us. He arrived one morning with his brother, also a doctor, two detectives, and half a dozen newspaper and camera men. I met them at the entrance.

"You haven't much time, Doctor," I said. "What would you most like to see?"

"Whales," he answered without a moment's hesitation. "I read a book of yours about whales and I'm much interested in how they are born."

The words were hardly out of his mouth before the news hawks were on the job. Oh boy, what a story! Quintuplets and whales! But the doctor was really serious. He never thought it would make a good story; he wasn't that sort of doctor. By chance he had come across my book *Whale Hunting with Gun and Camera*. In it I told a lot about baby whales and it made him very curious. We walked out to the Hall of Ocean Life mostly filled with specimens I had collected.

"First," said the doctor, "I want to know how big a whale is at birth."

"Well, that depends upon the species and the size of the mother. I took a baby thirty feet long, weighing about eight tons from an eighty-foot blue whale."

"My goodness! Eight tons, did you say? That's sixteen thousand pounds. My goodness! I never thought they were as big as that."

His questions were all about whales, to the exclusion of other subjects. When his visit ended and we were at the Roosevelt Memorial entrance I asked about the famous quintuplets.

"Well," he said, "I don't deserve any particular credit for just delivering them. But keeping them alive in the unsanitary conditions wasn't easy. The organizing afterward, that was the real job. And then, too, there were religious difficulties. But I've been there a long time and they sort of trust me."

With a deprecating laugh he dismissed the matter. A fine type, the doctor. He wasn't swept off his feet by the reception New York City had given him, either. He knew just what value to put upon it. He was enjoying it thoroughly but his wise gray head had not been turned one bit. A simple, kindly man, devoted to his work and the people among whom he lived. I don't wonder that "they sort of trusted" him.

The Crown Prince of Sweden came and the Duke of Windsor; dozens of world celebrities. We got them inevitably after Al Smith and the Empire State Building.

One extra-museum interlude was very interesting. A radio program had been arranged at the National Broadcasting Company called the "Order of Adventurers" with Admiral Byrd, Colonel (now General) Theodore Roosevelt, Lowell Thomas, Captain Felix Reisenberg, the explorer-author, and myself. We were all friends and the broadcast once a week gave us an excuse to see each other more frequently than we otherwise would have done. Not only did we tell stories ourselves but people who had had interesting adventures from all over the country were brought to the microphone and their experiences dramatized.

I never will forget the lady who had taken the gorilla "Gargantua" as a baby and reared him in her home. She told how one night during a thunderstorm, which always terrified him, she waked to find the beast in her bedroom. Had she made a single false move the gorilla would have killed her. She got quietly out of bed and with soothing phrases took the great ape by the hand, led him downstairs to the ice box, got a paper bag, and tossed it into the cage. He followed and she slammed the door. What was in the paper bag? Cream puffs!

Vincent Sheean told a story of the Riffs in North Africa. I hadn't seen Jimmy Sheean since he turned up at my Peking house in 1927, after making his way overland from the Yangtze Valley during the Russian attempt to communize China. Then, he was bedraggled and unkempt. That night he appeared immaculate in evening clothes cut by a London tailor. Jimmy never had seen a radio dramatization. During rehearsal an actress was supposed to be choking from the gases of a volcano. She did it so realistically that Jimmy thought she was actually choking, rushed for a glass of water, and pounded her on the back.

It was all great fun. I wish it might have continued but Dick Byrd started for the Antarctic on a new expedition. Felix succumbed to a heart attack, Lowell had other commitments; only Ted and I were left. So the program died a natural death.

Radio work is fascinating to me. For five years I've done the Wednesday program "New Horizons" on Columbia's "School of the Air of the Americas." Since nervousness in front of a microphone is gone I can really enjoy the dramatization. The sound effects man with his ingenious gadgets for simulating every known type of sound; the actors who do their part as though before an audience on the stage; the exact timing; all of it seems new and refreshing at every program.

In 1941, Trubee Davison went to Washington as aide to General Emmons in the Air Corps. Although he had been Assistant Secretary of War for Air during the Hoover Administration he wanted a job with active fliers. The board of trustees gave him leave of absence as president. That was a sad blow to me personally. I had become increasingly unhappy in the Museum. Raising money, trying to make both ends meet in our budget was the be-all and end-all of my existence. I hardly had time to walk through the halls. Planning new exhibits was useless for there wasn't a dollar to spend unless I could persuade someone to give us money. My work might almost as well have been carried on from an office in Wall Street. When Trubee left, the one bright spot in my Museum work "blacked out."

As I said before, it was no one's fault. We were the victims of war and taxes along with other institutions of a similar kind. The Museum was overextended in buildings. The situation was like that of a man who in 1929 had erected a big house. Then came the Depression. His income was not sufficient to maintain an adequate staff of servants, to do any entertaining, or even keep the house in repair. No board of trustees ever worked harder to solve the problems of an institution for which they were responsible than did ours. Curtailment of staff and activities was the only solution. Neither by temperament nor training was I the man to carry the Museum through this period of its existence. I was "a square peg in a round hole."

Billie felt that in justice to the institution and myself I ought to resign. I knew she was right but I couldn't quite make up my mind to leave active work in the Museum where I had spent thirty-five happy years. When it became evident that meeting the next budget would mean wholesale retirements of men like Frank Chapman, Clark Wissler, Barnum Brown, and others who had made the institution great, but were far along in years, I couldn't face it. So I offered my resignation to the board to take effect January 1, 1942, and was appointed honorary director.

It was high time. I had been brought up in the old Museum traditions. Everything had changed. Nothing in our lives would be the same again. All educational institutions must face radically different problems, must adapt themselves to a "new order." A younger man able to adjust himself and his ideas to the aftermath of the war was needed to direct the destinies of what I believe is one of the most worth-while institutions in all the world.

I was deeply sorry to leave the associations and day by day companionship of my scientific colleagues and the trustees. All of them were personal friends; some I had known since boyhood. Inevitably our paths would lead in different directions even though I came often to the Museum which I shall always love. Their interests lay in New York; mine were directed to the forests and lakes of the Berkshire Hills. It was the close of the longest chapter of my life.

The transition from active work in the Museum to Pondwood Farm was wonderful. Billie, with her feminine foresight, had been preparing the way for nearly two years. She knew that it was inevitable. When I came home and told her I was through as director she looked into my eyes for a long moment. Then her face broke into a radiant smile for she saw the relief she hoped to find. "Thank God you are free once more." My son George was there on leave from the Seventh Regiment. Both of them kissed me and we drank to a new episode of our life together.

Chapter 30

Berkshire Paradise

It was a beautiful day in the spring of 1942 in the Berkshires. Sun shining, birds singing. I was mooning about the orchard sniffing the buds of apple blossoms almost ready to burst into bloom. Billie called to me from the gun room.

"Roy, you've got to go over to Roberts's and buy some chickens. Stop loafing out there. You're a farmer now, not a museum director. And when you come back I wish you'd rake the leaves away from the stone wall and I want my wisteria vine tied up and you can take the branches off the rock garden and the burlap away from the shrubs and then, if you have any time left, I wish you'd see about a load of manure. I want sheep manure for my flowers and don't let Dave Curtiss tell you anything different. Isn't it wonderful to be here in the sunlight? And we don't have to go to New York and worry about museum budgets any more. Oh, I'm so happy!"

And so we were and so we are. The release from official duties, to live again in the sunshine and the open air; to write when I want to write, to work in the good earth, to fish and shoot and swim, is Paradise enough. I feel as I did the night the *Albatross* rode into Keelung harbor, after battling a typhoon in the Formosa channel. My personal ship during thirty-five years has sailed all the oceans of the world; now at the end of its restless voyaging it has come to anchor in the quiet waters of Pondwood Pond.

Life, for me, has been a series of distinctly separate episodes and violent contrasts. Each one, apparently, had no relation to the immediate past, yet, in reality, it was a logical outgrowth of what went before. I had written ten books and dozens of magazine articles, but never thought of myself as an "author." Every explorer must tell what he has done in order to maintain public interest and support. Writing was a necessary corollary of exploration.

Suddenly the position was reversed. What had been an avocation became my vocation. Literary interests were paramount and Billie's and my personal contacts changed immediately. Men and women of New York's social world, scientists and educators and Wall Street financiers gave place to publishers and writers, editors and artists. It was like turning off the light in one room and switching it on in another. At the Dutch Treat Club luncheons I found a concentration of the men who make public opinion through magazines and newspapers, radio, illustrations, and books. For many years, I had been a member of the club. I went to the Tuesday luncheons when possible but often missed them because of Museum duties. Now I am always there.

Two hundred or more of us gather every week at the Park Lane Hotel. At the small table where I usually sit are Arthur Train the novelist, Bill Chenery of *Collier's,* Will Beebe, Percy Waxman of *Cosmopolitan,* John Kieran, the savant of Information Please, Sumner Blossom, who edits the *American Magazine.* Lowell Thomas may sit with us, or Roy Howard or Clarence Budington Kelland. One can look about the room and see a hundred men whose names stand high in the "creative arts." Richard Crooks, John Charles Thomas, Reginald Werrenrath of the Metropolitan Opera Company; John Golden, ex-President Hoover, Hugh Gibson, Ray Vir Den, Burns Mantle, Deems Taylor, Grantland Rice, Rube Goldberg, John Erskine, Westbrook Pegler, and so on *ad infinitum.* The men and women who make the world's news come every week to our speaker's table. It is a "Reader's Digest" of current events. Last year I was elected to the "Inner Sanctum," the board of governors of the club. Never has an honor been more appreciated.

During the winter, Billie and I go to New York two days each week; I for the Dutch Treat luncheon and my radio program, she to shop, visit her especial friends, or to do whatever is on the cards. But in the summer the city sees us no more for our life centers at Pondwood Farm. It offers something interesting every day; there is not a moment of idleness. Billie has charge of the gardens. An acre grows vegetables of every conceivable kind and her flowers take a lot of time besides the multiple interests of the house. As for me, I fish early in the morning and then retire to my log cabin in the forest to write for five or six hours, or work about the place. At cocktail time, Billie and I sit in the rock garden facing the pond. We discuss what each has contributed to "Andrews Incorporated" during the day. Perhaps I read the beautiful thoughts I have set down on paper. They may sound beautiful to her or not—she always tells me frankly.

At Pondwood Farm, we have a balanced program of work and play. Sport is just as important as the serious side of life. Guests are often with us. Reg Rowland and his lovely Beatrice; Frederick Barbour and Helen; Abel I. Smith and Dottie; Annette and Ashley Howe; Kay and Rodney Williams or Fred Walcott. These are our especial friends because we all have similar interests.

Billie can kill a woodcock or cast a fly as well as I and she loves it, too. The other wives are not so keen as she but all of them look with an indulgent, if somewhat superior, eye on the childlike enthusiasm of their husbands.

While we sit in the rock garden sipping cocktails, suddenly the rainbow trout or bass begin to rise. The pond is dimpled with a dozen ripples and fish leap above the white pond lilies. The men rush to the gun room for rods. We take our cocktails with us in the boats for the trout and bass stop feeding on the surface as suddenly as they begin. You catch them then or not at all. Many of the bass reach four pounds or more and few of the trout are less than two pounds else they couldn't hold their own with the big mouth bass. Of course, I stocked the pond. The bass have done amazingly well and rear their families by the score. But the trout will not breed without a running stream and I replenish them every year.

The woodcock and duck season opens October fifteenth. For weeks before, we have planned every moment of the day. Reg Rowland, Frederick Barbour, and I have worked our bird dogs since early in September. Reg is one of the best wing shots I have ever known. He *thinks* like a woodcock or grouse and knows where they will be found and what they will do under given conditions; therefore he gets more birds than any of us. The opening day is the great moment of the year. The duck blinds on Smith Pond have all been assigned; each of us knows what woodcock covers the others will shoot. We go to bed early, filled with excitement.

Long before daylight the alarm clock rings. Billie snuggles her head deeper into the pillow. "Why should anyone get up at this ungodly hour?" she murmurs. "You shoot your old ducks. I'm not going."

"All right, darling," I say, knowing full well she wouldn't miss it for the world. "You go back to sleep. I'll go alone." So I start to dress. That does the trick. In a few minutes she says crossly, "I just don't understand why ducks can't be shot at a human hour. But I'm awake now. Go put the coffee on."

So I start the coffee. When I return she is pulling on hip boots and sweater. Her eyes are shining. She runs to the window. "I can see a smitch of daylight already. Why didn't you set the clock earlier? We'll never get there in time. Reg and Frederick and Rodney will beat us."

But we do get there and are snuggled in the blind as the first gray light of dawn begins to show the decoys bobbing on the water like black shadows. We have to wait till sunrise but the ducks come in. We argue in whispers whether or not certain bright sky streaks can legally be called "sun-up." We decide they can and then the fun begins. Billie is a fine shot. She didn't know one end of a gun from the other until she married me, but practice and perfect natural co-ordination have made her shoot as well as the average man.

About half past eight, the first act of the opening day is ended. We go home for breakfast and then with our dogs into the woodcock covers. I wonder how many people who read this book know the sort of place in which the "timber doodle" lives! It is an alder swamp or a scrub-birch hillside so thick you can hardly see ten feet. Queen and Star wear bells; otherwise they'd be lost in a moment. We work slowly through the smother, pushing alder saplings aside. I hear a cry from Billie. "Oh, that branch hit me right in the face. Oh, oh, it hurts so. *Damn* the woodcock." She wipes tears from her eyes, gets her gun up, and pushes on. "I can't hear Queen's bell. She must be on a bird. Yes, there she is. Right in front of me. She's pointing." We find Queen, stiff as a ramrod, her lips trembling. I take a line from her nose. "The bird is right over there. You work around to that little open space. I'll go in on the dog. Ready?"

I move in and the woodcock flutters up like a bat in daylight. Bang, goes Billie's gun. "You got him. Good girl. Damned fine shot." Queen brings back the bird.

At the end of the morning we go home. I telephone the other men to find what their score has been. We haven't done so badly, either on the ducks or woodcock; in fact we're well up front. After lunch, an hour's sleep is indicated for we've been up since before dawn. Then the afternoon shoot. Last year at six o'clock in the afternoon of the first day I came into our bedroom. There on the bed lay Billie with one arm extended, pillowing Queen's black and white head which was half covered by Billie's blonde curls. Both sound asleep. The end of a perfect day!

At seven on the opening evening our friends gather at Pondwood Farm for a buffet supper. We men are all in shooting clothes just as we came from the field. Every hour, every shot, how each dog worked is told and retold. Such are our days of sport when the leaves have turned to red and gold and the harvest moon glows in the sky.

There are other sides to our life at Pondwood Farm. Good neighbors. What does the word mean in the city? Nothing. I have lived in a New York apartment for months without knowing the name of the people above or below or on either side of me. Franklin Roosevelt has made the term almost a bromide but it becomes something vital and important when one lives in the country. Last winter, during ice storms and blizzards, twice we were completely isolated. No lights, heat, water, or telephone; drifts eight feet high across our road. Storms such as Connecticut had not seen in a generation cut us off from the outside world. I was chained to a couch with a broken leg. Neighbors fought their way to our house to give what aid they could. They came with snow plows, firewood, and food. Just because they were good neighbors!

Another gift to us of Pondwood Farm was a new and different life with my son George. With only an apartment in town when he was at St. Paul's and Princeton there were the multiple attractions of New York during school vacations. But the farm changed all that.

After George went in the Air Corps there was one wonderful week for me when we shot woodcock and grouse together over Queen and Star. We sat on a log, smoking our pipes, talking of his future and mine.

"Dad," said he, "I'd just love to come to Pondwood Farm and be a country squire."

But he won't. The restless spirit, the inheritance, which makes him delight in the freedom of the skies will call him to the far places of the earth even as it did his father.

What, you may well ask, has been our part in this world war? By necessity it is not important. I offered my services for active duty. Too old. At fifty-nine I am physically fit, but this, they say, is a young man's war. No one can guess what next year will bring. Perhaps they will find a place for me in the Orient which I know so well.

Such is a kaleidoscopic picture of my life up to the Year of Our Lord, 1943, the first day of the month of March. I loved living it and it has been almost as much fun to write the story. Always there has been an adventure just around the corner—and the world is still full of corners!

Afterword: The Andrews Legacy

by Ann Bausum

Roy Chapman Andrews applied twin talents that yielded remarkable results in his scientific career both during his lifetime and beyond. Neither his fame nor his legacy would be as notable without his strategic blending of these two skills: storytelling and leadership. As raconteur and leader, Andrews raised money, promoted his research, inspired budding scientists, and ensured opportunities for field research—and results—as Charles Gallenkamp observes in this volume's Foreword.

Andrews recalled the challenges and thrills of his lifetime of scientific adventure in a number of accounts, from lectures to magazine stories to books. None of these summarizes his career better than *Under a Lucky Star.* Published in 1943 by Viking and long out of print, the book is finally available again in this new, authorized edition.

The title of this most comprehensive of Andrews autobiographies reflects the explorer's realization that he had lived his life "under a lucky star." In the early pages he recounts a youthful misadventure, a boating accident in which he almost drowned and his companion perished. The guilt-racked Andrews questioned why his friend, a young professor at Beloit College and "the good swimmer, was the one to die almost within reach of safety while I survived against well nigh impossible odds. Still, it happened that way several times later in my life," he adds. Thus began the twinkling of a lucky star that would see Andrews safely through sandstorms, tropical illness, and a swarm of pit vipers, among other assorted perils.

In 1926, Andrews tallied an early list of the star's credits with his typical flair for humor and understatement. "In the [first] fifteen years [of field work] I can remember just ten times when I had really narrow escapes from death," he wrote in *On the Trail of Ancient Man*. Then he unfurled his list: "Two were from drowning in typhoons, one was when our boat was charged by a wounded whale; once my wife and I were nearly eaten by wild dogs, once we were in great danger from fanatical lama priests; two were close calls when I fell over cliffs, once I was nearly caught by a huge python, and twice I might have been killed by bandits." Thank goodness for that lucky star!

Southern Wisconsin served as both birthplace and training ground for Andrews's lifetime of adventure and discovery. He was born on January 26, 1884, in Beloit, along the Wisconsin-Illinois border. His hometown harbored a liberal arts college and industry amid rich farmlands, prairies, and woods, and his home at 419 St. Lawrence Avenue then rested on the western edge of the city. (It remains in use as a private residence.) He exhibited the calling of a naturalist from an early age. "I was like a rabbit, happy only when I could run out of doors," he recalls in *Under a Lucky Star*. "To stay in the house was torture to me then, and it has been ever since."

Nothing seemed more natural during that era than for a boy growing up in rural America to become acquainted with firearms. When Andrews received a shotgun on his ninth birthday, he began steps on a path that would lead, almost fatefully, to a lifetime of exploration. The young boy prowled the neighboring prairies, stalked prey in the area's open-oak grassy woodlands, and navigated its local waterways—armed all the while with the essential tools of a nineteenth-century naturalist: a gun and a notebook.

Andrews proved to be an excellent marksman. He taught himself taxidermy and established a makeshift natural history museum filled with stuffed birds and other collections in the barn behind the family home. He read extensively, began to dream about becoming an explorer, and developed an admiration for the American Museum of Natural History in New York City. His skills as a taxidermist had the added practical dimension of generating income. "Every bird and deer head shot within a radius of fifty miles came to me if a sportsman wanted it mounted," Andrews boasts.

School became a necessary evil; it offered new windows for knowledge but charged the heavy price of indoor instruction. Andrews became a selective student, performing well in courses he liked, such as English, and struggling through those he did not, such as math. He enrolled as a day student at Beloit College, saving the cost of room and board by walking across town and over the Rock River each day. His years of study coincided with the growth of the school's Logan Museum of Anthropology and the offering of its first courses in evolution. Only the previously mentioned boating accident, which occurred in 1905 during his junior year, marred his college experience.

After graduating the following year, Andrews acted on his ambition to become an explorer and talked his way into a job at the American Museum of Natural History. "I just want to work here," he pleaded with the director after traveling to the city without any guarantee of finding work. Andrews's persistence landed him a job in the taxidermy department where he literally started at the ground level, mop in hand. (He would later become the museum's director.) The career that followed fills the pages of this book and yielded scientific riches that continue to be studied and displayed today. To Andrews's dismay, however, events beyond his control brought his exploration career to a premature end. Not even a lucky star could overcome the combined obstacles of political turmoil in Asia and a worldwide economic depression. He left the field reluctantly in 1932, at age forty-eight, never to return.

Although Andrews never again lived in Wisconsin after graduating from Beloit College, he maintained his connections to the area. For example, his 1928 expedition staff included Alonzo

Pond, a native of nearby Janesville, a 1918 graduate of Beloit College, and archaeologist at the Logan Museum of Anthropology. Andrews shared collections from this expedition with the Logan Museum, most notably Paleolithic stone tools from Inner Mongolia. His alma mater awarded him an honorary degree in 1928. He continued to visit his hometown as long as his parents lived there. Most significant of all, perhaps, the intrepid explorer arranged to make one final trip home: after his death from a heart attack in 1960 at age seventy-six, his ashes were interred in the family plot at Beloit's Oakwood Cemetery.

Andrews, the equal in stature and fame during his lifetime of explorers like Byrd, Peary, and Shackleton, might have faded into obscurity had it not been for three events. The first was the publication of Douglas Preston's book *Dinosaurs in the Attic*. This 1986 history of the American Museum of Natural History placed Andrews front and center among the museum's most noted explorers. The illustrated chapter on Andrews, "Fossils in Outer Mongolia," helped restore his forgotten reputation. Another event was Hollywood's creation of the daring Indiana Jones. Persistent rumors credit Andrews with serving as the real-life model for the adventurous lifestyle of the imaginary explorer Dr. Henry Jones, Jr.

Andrews himself planted the seeds for a third boost to his reputation by devoting the final years of his life to storytelling and writing. Arguably his most influential books were the three volumes he contributed to the Random House "All About" series for children: *All About Dinosaurs, All About Whales,* and *All About Strange Beasts of the Past.* A surprising number of young people who read these books grew up to pursue careers in science. Among those who speak fondly of their influence is Michael Novacek, a paleontologist whose career has likewise taken him to the American Museum of Natural History (curator of paleontology, provost of science, and senior vice president) and into the Gobi. Novacek helped reestablish exploration by Americans in Mongolia following a seventy-year hiatus brought on by Andrews's departure from the region in 1930.

The revival of Andrews's reputation is reflected by his increased visibility in the exhibits of the American Museum of Natural History, the growing number of books about him, and his presence in a series of scientific videos and films. His latest and most significant appearance is in an IMAX production called *Dinosaurs Alive,* released in 2007, which pairs extensive clips of film footage from Andrews's own Gobi expeditions with contemporary views of modern explorations there by American Museum of Natural History scientists, including Michael Novacek.

Even as the appreciation of Andrews has gained traction nationally, interest in him has grown again in his hometown. In 2000, forty years after his death, Beloit-area citizens founded the Roy Chapman Andrews Society to honor his legacy as an explorer and as a role model for young people with an interest in science. The group sponsors a website, www.RoyChapmanAndrewsSociety.org, about Andrews, helps promote local points of interest related to his life, and offers programs that honor this most famous hometown hero. The cornerstone of this effort is the society's Distinguished Explorer Award. This annual recognition honors contemporary explorers with a commemorative statue and cash prize during a public celebration on the campus of Andrews's alma mater. The central aim is to introduce local schoolchildren to modern explorers in the hopes of inspiring yet another generation of young people to carry on the Andrews legacy.

Ann Bausum is the author of Dragon Bones and Dinosaur Eggs: A Photobiography of Explorer Roy Chapman Andrews *and several other books published by the National Geographic Society. She lives in the hometown of Roy Chapman Andrews, Beloit, Wisconsin.*

Colophon

This book is set in Sabon, an old-style serif typeface designed by the German-born typographer and designer Jan Tschichold (1902–1974). It was designed by Ken Crocker and printed and bound by Worzalla Printing.